MAMMOTHS, SABERTOOTHS, AND HOMINIDS

MAMMOTHS, SABERTOOTHS, AND HOMINIDS

65 MILLION YEARS OF MAMMALIAN EVOLUTION IN EUROPE

JORDI AGUSTÍ and MAURICIO ANTÓN

COLUMBIA UNIVERSITY PRESS
New York

Columbia University Press
Publishers Since 1893
New York   Chichester, West Sussex

Library of Congress Cataloging-in-Publication Data

Agustí, Jordi, 1954–
    Mammoths, sabertooths, and hominids : 65 million years of mammalian
evolution in Europe / Jordi Augustí and Mauricio Antón.
        p.   cm.
    Includes index.
    ISBN 0-231-11640-3 (acid-free paper)
    1. Mammals, Fossil—Europe. 2. Paleontology—Tertiary. 3. Animals, Fossil—
Europe. I. Antón, Mauricio. II. Title.
    QE881 .A35 2002
    569′.094—dc21   2001042251

Columbia University Press books are printed on
permanent and durable acid-free paper.
Printed in the United States of America
c 10 9 8 7 6 5 4 3 2 1

CONTENTS

Preface **vii**

CHAPTER 1   THE PALEOCENE: THE DARK EPOCH • **1**

CHAPTER 2   THE EOCENE: REACHING THE CLIMAX • **23**

CHAPTER 3   THE OLIGOCENE: A TIME OF CHANGE • **67**

CHAPTER 4   THE EARLY TO MIDDLE MIOCENE: WHEN
THE CONTINENTS COLLIDE • **93**

CHAPTER 5   THE LATE MIOCENE: THE BEGINNING
OF THE CRISIS • **151**

CHAPTER 6   THE PLIOCENE: THE END OF A WORLD • **211**

CHAPTER 7   THE PLEISTOCENE: THE AGE
OF HUMANKIND • **241**

Bibliography **283**

Index **295**

PREFACE

T O SUCCEED IN WRITING A BOOK ABOUT THE AGE OF MAM-
mals in Europe appears to be a difficult goal, especially in light
of such brilliant precedents as *The Age of Mammals* by H. F. Os-
born and *The Age of Mammals* by Björn Kurten. In between this book
and the 1971 version of *The Age of Mammals* there is not only the extraordinary scientific personality of Kurten but also thirty years of additional knowledge about fossil mammals and the environment in which
they evolved. In the past decades, the amount of new information on
some selected time slices in the history of our planet has increased enormously. This is particularly true for the Pleistocene, but also for the
Miocene and Pliocene (the so-called Neogene); and even the oldest
periods of the Cenozoic, Paleocene, Eocene, and Oligocene are much
better known today than at the beginning of the 1970s. At the same
time, there are texts dealing with the subject outside Europe, such as
C. Janis, K. M. Scott, and L. L. Jacobs's *Evolution of Tertiary Mammals of
North America,* and others in preparation, such as A. Turner and M.
Anton's *Evolving Eden* on the faunas of African mammalian fossils. We
have, therefore, restricted the scope of our topic to Europe, in the wider
sense: from Iberia to the Urals and Georgia.

The increase in information over the past few decades has affected
this work in two ways. First, we have had to restrict the systematic and
phylogenetic aspects of each group we have included. Second, we have
had to adjust our level of chronological resolution, which has increased
significantly thanks to the application of new geological and geophysical
techniques such as paleomagnetism and cyclostratigraphy. We can now
distinguish on a scale of hundreds or dozens of thousands of years
events that hitherto appeared to be synchronous or closely related in
time. This has introduced a degree of complexity when planning this
book. The aim is to satisfy a wide range of readers, from someone who
is interested in the subject (perhaps a geologist or an anthropologist)

to the students or postgraduate scholars who wish to have an overall idea of the origins of European mammalian faunas without immersing themselves in the copious bibliography available on the subject. This forces us into relative simplification so that readers can understand the events that affected Europe over the past 65 million years. For example, nowadays we have a clear and detailed grasp of the sequence of events that took place during the first contact between the Eurasian and African plates during the early Miocene. We also have information about the climatic crisis that followed in the middle Miocene. However, the corresponding chapter is not structured event by event because readers would be lost in a multitude of successive immigrations, emigrations, and extinctions. Rather, we have preferred a consideration of the early Miocene as a whole, with a group-by-group analysis of what occurred during this time slice. Even so, this book must address the real complexity of the history of Europe during the Cenozoic. Europe, although just an appendage of the Eurasian supercontinent, acted during most of its history as a crossroad where Asian, African, and American faunas passed one another, throughout successive dispersal and extinction events.

But these events did not happen in an isolated context, since they were the response to climatic and environmental events of a higher order. Thus this book pays special attention to the abundant literature that for the past few decades has dedicated itself to the climatic evolution of our planet. Today more than ever, the explanation of any group's evolution appears as an element within the global working of the Earth system. The evolution of a group can, therefore, be interpreted as a new response to the several environmental crises that have affected the planet throughout its history. Compared with other periods in Earth's history, the Cenozoic was particularly complex in terms of climate, geography, and geodynamics. The change from the evergreen tropical forests of the Eocene to the freezing steppes of the late Pleistocene cannot be pictured as the result of a regular and gradual drop in temperature. On the contrary, from the first Antarctic glaciations to the successive climate optima of the Miocene and Pliocene, including such events as the drying up of the Parathethys and Mediterranean Seas, the geological and environmental history of Europe shows a series of fluctuations that have determined the evolution of terrestrial faunas. This book provides a general impression of this evolution. Despite this, readers will discover an underlying pattern to this whole history. Thanks to isotopic analyses developed from ocean drillings by researchers such as N. Shackleton, we now have a clear idea of the evolution of the paleotemperatures and the growth of ice at the poles. However, we still do not fully comprehend the way in which oceanic changes affect terrestrial ecosystems. Despite the innate desire of the paleontologist to demonstrate an exact correlation between changes in climate and changes at the level of terrestrial ecosystems, the question seems more complex, since large oceanic changes seem to have had only a limited effect at the continental level, while important restructuring at the level of terrestrial ecosystems seems to correspond to much smaller oceanic shifts.

So there is still much work to be done before we understand the true factors that have driven the evolution of terrestrial faunas. For this reason, in addition to information that refers exclusively to mammals, this work includes information on vegetation and invertebrates. In contrast, this book does not deal with major questions related to the concrete phylogeny of the families, genera, and species that are mentioned. Only when the origin of a genus has some relevance from the point of view of biogeography or paleoenvironment is specific mention of its origins (for example, the character of the fauna from Baccinello, during the late Miocene of Italy, changes radically if the hominoid *Oreopithecus* is related to some African anthropoid or descends directly from the European *Dryopithecus*). Readers who are interested in concrete phylogenetic questions can avail themselves of the extensive bibliography on the subject. Finally, readers will note that in this book we make almost constant reference to the genus as the taxonomic category and, to a lesser extent, the species. This is common to all works of this sort, which require a minimum of explanation. A genus is simply a grouping of species and can never be considered as an ecological or an evolutionary unit. However, its use in this work responds to the need to simplify, since for periods such as the Eocene and even the Miocene, reference to specific species within a genus would have complicated the reading of the text. The distinction between genus and species has been maintained, though, when it is of interest to refer to different species within a genus (perhaps because they corresponded to very different morphotypes, as was the case with the felid *Pseudaelurus*). In other cases—when we refer, for example, to *Dinocrocuta*—we are obviously referring to the group of individuals that make up the different species of this genus.

A final comment on the references and bibliography in this book: obviously, the vast quantity of information collected here comes from many different sources and references. To include them all in the book again would have converted it into a dense text. Thus the bibliography contains expressly all the works that have been used in the writing of this book. In particular, those works that have served as sources on the lifestyle and weight or size of each group are asterisked. For example, if readers want to know the source of information on the weight of ruminants or amphicyonids in the Miocene, they will see that an asterisk is beside the reference to Köhler (1993) or Viranta (1996), respectively. Most of the references on paleoclimatic or paleogeographic evolution are usually cited directly in the text because it would have been much more difficult for the reader to identify in the bibliography the concrete source for each climatic or paleoenvironmental datum.

ACKNOWLEDGMENTS

This book has considerably benefited from the comments, suggestions, discussions, advice, and encouragement of many friends and colleagues, including Luis Alcalá, José María Bermudez de Castro, Ray Bernor, Louis de Bonis, Harold Bryant, Angel Galobart, Léonard Ginsburg, Meike Köhler, Jan Van der Made, Jorge Morales, Salvador Moyà, Jerome Quiles, Lorenzo Rook, Benedetto Sala, Bernard Sigé, Jean-Pierre Suc, and Alan Turner. We also thank the Institute of Paleontology M. Crusafont (Sabadell) for allowing us to reproduce plates 1, 2, 3, 6, 8, and 12.

We thank Anthony Chiffolo and, at Columbia University Press, Irene Pavitt for making this book more readable and more accessible to a general audience. We also acknowledge Alessandro Angelini and, for his support of this project, Robin Smith, both of Columbia University Press.

Finally, during all this time we have had the support and patience of our wives, Esperanza and Puri. This book is dedicated to them.

MAMMOTHS, SABERTOOTHS, AND HOMINIDS

CHAPTER 1
The Paleocene: The Dark Epoch

A COMMON SCENARIO TENDS TO POSIT THE EARLY EVOLUTION-
ary radiation of placental mammals as occurring only after the
extinction of the dinosaurs at the end of the Cretaceous period.
The same scenario assumes a sudden explosion of forms immediately
after the End Cretaceous Mass Extinction, filling the vacancies left by
the vanished reptilian faunas. But a close inspection of the first epoch
of the Cenozoic provides quite a different picture: the "explosion" began well before the end of the Cretaceous period and was not sudden,
but lasted millions of years throughout the first division of the Cenozoic
era, the Paleocene epoch. Following the partition of the Tertiary period
by Charles Lyell into the Eocene, Miocene, and Pliocene epochs, the
paleobotanist W. P. Schimper added the Paleocene in 1874 to include
a number of fossil floras from the Paris Basin that preceded the first
Eocene levels. These Paleocene strata were characterized by the presence of primitive mammals preceding those of the Eocene epoch, in
which the first members of the modern orders were already recognizable (perissodactyls, artiodactyls, rodents, bats, and others).

The term *dark* has been applied to the Paleocene epoch in the title
of this chapter for two main reasons. First, our knowledge of this remote
time of mammalian evolution is much more obscure and incomplete
than our understanding of the other periods of the Cenozoic. Second,
compared with our present world, and in contrast to the succeeding
epochs, the Paleocene appears to us as a strange time, in which the
present orders of mammals were absent or can hardly be distinguished:
no rodents, no perissodactyls, no artiodactyls, bizarre noncarnivorous
carnivorans. In other words, although the Paleocene was mammalian
in character, we do not recognize it as a clear part of our own world; it
looks more like an impoverished extension of the late Cretaceous world
than the seed of the present Age of Mammals. But the seeds were there.

The Aftermath of a Mass Extinction

Spanning no more than 10 million years, between 65.5 and 55.5 million years ago, the Paleocene began after the great event at the end of the Cretaceous that ended more than 150 million years of reptilian domain over the continents and seas. Despite its catastrophic biotic effects, this event seems not to have deeply affected the long-term, overall conditions of the planet. On a global scale, there appear to have been few variations between the late Cretaceous and the early Paleocene climates. The analysis of plant distribution in Canada immediately after the crisis supports this statement, indicating a quick recovery of species of palms and screw pine, in proportions similar to those existing now in Southeast Asia (Nichols et al. 1986).

The distribution of landmasses during the Paleocene was quite different from the arrangement of land today. From the breaking away of the supercontinent Pangaea at the beginning of the Triassic, new marine domains opened throughout the Mesozoic, such as the Atlantic and Indian Oceans. At the beginning of the Paleocene, there were still some remnants of the broken Gondwana, the southern Pangaean continent. Thus Australia and South America were still attached to Antarctica (which was about where it is today). Africa and India, the other original Gondwanan "fragments," were far from both these southern continents and their present position close to Eurasia. At this point in the Paleocene, they "floated" isolated from the other continental landmasses, surrounded by the Atlantic and Indian Oceans and the Tethys Sea. The Tethys Sea formed an east–west continuous marine belt that separated North America from South America and Eurasia from Africa and India. This means that there was a continuous equatorial current, from the eastern to the western Pacific, throughout the Indian and Atlantic Oceans and the Tethys Sea. Along the coasts irrigated by this warm current, tropical reefs and corals recuperated after their near demise during the End Cretaceous Mass Extinction. Giant protists called alveolines, in symbiosis with green microscopic algae, formed massive banks of limestone, now embodied and pushed up into mountain ranges such as the Pyrenees and the Alps as a consequence of the Alpine orogeny. To the north of this Tethyan equatorial current, North America and Europe were still connected through Greenland, as shown by the similarity of the Paleocene and early Eocene mammalian faunas of both continents. North America was also connected to eastern Asia through the Bering corridor. In contrast, Europe and Asia were separated during most of the Paleocene and Eocene by a shallow sea that extended through the eastern margin of the Urals (the Turgai Strait) and connected the Arctic Ocean to the Tethys Sea.

Temperatures at the beginning of the Paleocene were 2 or 3°C lower than those of the late Cretaceous. From the early Paleocene to the middle Eocene, the average surface temperature of the oceans underwent a gradual increase, reaching levels between 2 and 4°C warmer than today. However, evidence indicates that during the Paleocene, temperatures fluctuated from cooler to warmer and again to cooler by the end of this epoch. The climate was humid, and the dominance of arboreal taxa indicates the extension of canopy rainforests over most of the continents. There is also geological evidence for the existence of dry–wet fluctuations, as shown by the huge amounts of gypsum and salts that were deposited subsequent to the desiccation of ancient lakes and basins, such as those of Ager and Tremp to the south of the Pyrenees.

The Cretaceous Inheritance
of the Paleocene

The first evolutionary radiation that led to Paleocene mammalian diversity began well before the end of the Cretaceous, and by the late Cretaceous a varied fauna of placental mammals already existed, including several insectivores and some primitive ungulate and primate-like species. These eutherian taxa joined other previously successful mammalian groups, such as the multituberculates and the marsupials. With the exception of this last group, which was severely affected by the End Cretaceous Mass Extinction (they declined from nine to one marsupial genus), all the remaining groups traversed this boundary without major variations. Therefore, the basal Paleocene mammalian faunas were not so different from those of the late Cretaceous and were basically composed of a number of families rooted in the Mesozoic. Among them, the most successful and diversified one was the multituberculates.

The multituberculates were a peculiar group of primitive mammals whose origins can be traced back to the early Jurassic and, perhaps, even the late Triassic. They have been so far the most successful and long-lasting order of mammals, having survived for more than 100 million years until their extinction during the Oligocene. They were not truly therian mammals and were probably closer in biological terms to the living monotremes than to marsupials and placentals. The cranial and dental anatomies of multituberculates looked like those of rodents, with long, chisel-like incisors separated from the cheek-teeth by a large space without teeth (diasteme). The term *multituberculate* refers to their peculiar dental morphology, each cheek-tooth displaying a number of parallel rows of small cusps that operated against a similar counter-row in the upper or lower jaw (*multituberculate* means "several cusps"). Altogether, the masticatory apparatus of the multituberculates formed, as in present rodents, an efficient chopping device.

From this basic rodentlike design, the multituberculates radiated into a variety of morphotypes during the Cretaceous and the Paleocene. Some of them, like the ptilodonts, were squirrel-like arboreal forms. The most outstanding feature of the ptilodonts was the peculiar shape of their last lower premolar, larger and much more elongated than the other cheek-teeth, its occlusive surface forming a serrated slicing blade. This was perhaps used for crushing and opening hard seeds and nuts. However, most of the small multituberculates like the ptilodonts probably supplemented their diet with insects, worms, and fruit.

Thanks to the well-preserved specimens of the ptilodont *Ptilodus* recovered from the Bighorn Basin in Wyoming, we know that these multituberculates could abduct and adduct their hallux and had the foot mobility characteristic of such arboreal mammals as present-day squirrels, which descend trees headfirst.

A group of European multituberculates developed a dental design like that of the ptilodonts, with an enlarged bladelike lower premolar. These were the kogaionids, first known from the late Cretaceous beds of Hateg in Romania. The most successful genus of this family, *Hainina*, was once thought to be a ptilodont because of its large, bladelike premolar in the lower jaw. However, further analysis has shown that *Hainina* and its late Cretaceous ancestor *Kogaionon* were primitive multituberculates, with molars bearing a smaller number of cusps and retaining a fifth premolar, a characteristic relating them to some Jurassic genera and not to the late Cretaceous ptilodonts. This unique combination of archaic and advanced characteristics indicates that a large, bladelike premolar was independently acquired at least twice during the late Cretaceous, by both the North American ptilodonts and the European kogaionids.

The taeniolabids were quite different from *Hainina* and the ptilodonts. They were a group of multituberculates that, unlike the latter, had a much heavier, more massive anatomy. They reached the size of a beaver and probably had a fully terrestrial lifestyle. While the ptilodonts succeeded in North America and the kogaionids in Europe, the highest taeniolabid diversity was found during the late Cretaceous and Paleocene in Asia, suggesting that taeniolabids originated there.

Besides the multituberculates, a diversified fauna of relatively nonspecialized placental mammals existed at the end of the Cretaceous and persisted into the Paleocene. Most of them were originally included in the order Insectivora because of their archaic dentition, which indicates insectivorous habits. Insectivorousness requires the crushing of hard but fragile exoskeletons, made possible with individual tapering points rather than long blades, as in the carnivores. However, other than sharing a number of archaic, nonspecialized features, there is little positive evidence for assembling these archaic groups into a natural phylogenetic category. They took part in the first placental evolutionary radiation during the late Cretaceous, which continued during the Paleocene.

The leptictids best exemplified the characteristics of this group. The leptictids were archaic "insectivorous" placental mammals that originated during the late Cretaceous and became extinct during the Oligocene. Their cranial and dental anatomy was so archaic and basic for a placental mammal that establishing their close relationship to other groups is difficult. Most of the postcranial anatomy and the lifestyle of this group have been inferred from the best preserved specimens of *Leptictidium* from the middle Eocene of Messel, Germany (figure 2.7). According to this evidence, the leptictids were small placentals, with a body length between 60 and 90 cm, that bore a complete, archaic dentition including incisors (two or three), canines (one), and V-shaped premolars (four) and molars (three). The head ended in a long, slender snout that probably displayed a short trunk. This trunk likely was used for scratching in the undergrowth in search of insects and worms. According to the middle Eocene specimens of *Leptictidium,* the forelegs were extremely shortened, while the hind legs were elongated. This suggests a kind of locomotion similar to that of small kangaroos or jerboas, which use their elongated hind limbs for jumping. However, the leptictids' tarsal anatomy contradicts this supposition, indicating a specialization for running on the ground. Most probably, they were capable of both kinds of locomotion, running slowly on the ground in search of food and jumping quickly in case of danger. A surprising feature of the Messel specimens is the extraordinarily long tail, formed of about forty vertebrae—a unique feature among modern placental mammals. They probably used the long tail for balance while jumping or running quickly.

Like the leptictids, the palaeoryctids are known from the late Cretaceous of North America and had a generalized placental anatomy. From a nearly complete skull of *Palaeoryctes* from the Paleocene of New Mexico, we know that they were probably small, shrewlike insectivores with a long snout like that of the leptictids. In contrast to our knowledge of that of the leptictids, we know little about the palaeoryctids' postcranial anatomy. Unlike the short-lived leptictids, the palaeoryctids seem to have been ancestors of one of the groups that was going to succeed in the Eocene. Thus although the dentition of the palaeoryctids still indicates a mainly insectivorous diet, some details of their dental morphology relate them to the creodonts, the carnivorous order that filled the predator guild during the Eocene.

Another archaic, "insectivorous" placental group was the pantolestids. As with the leptictids, the best evidence of the pantolestids' full anatomy and lifestyle comes from the beautifully preserved specimens of the middle Eocene of Messel. According to the data provided by *Buxolestes* from this locality, as well as other, less complete specimens, the pantolestids were semiaquatic fish predators with a body of about 50 cm that ended in a long tail of about 35 cm. They bore moderately strong canines and multicusped cutting teeth, which were supported by a strong jaw musculature. The forearms were powerful and ended in large bony claws. The ulna and the radius were free, allowing wide rotational movement. Both features probably indicate its ability to dig and build underground dens. The hind limbs were also powerful but could not be rotated in the same way as the forelimbs. The first vertebrae of the tail presented strong transverse expansions, or processes—which suggests that the tail was moved powerfully in the water, with movements reminiscent of those of other semiaquatic mammals like otters.

A fourth group of small archaic placentals that are often included among the insectivores was the apatemyids. Unlike the leptictids and pantolestids, the apatemyids were rather specialized forms bearing a complex dentition. In relation to their relatively small body, they had a large skull armed with two extremely large, curved incisors. These operated against a similar pair of strong procumbent incisors in the lower jaw and were followed by a serrated, scissorlike battery of premolars. In contrast to this specialized frontal dentition, the molars were relatively small. From the evidence available from another middle Eocene genus from Messel, we know that the apatemyids used their sophisticated dentition to tear open the bark of wood in their search for insect larvae. The apatemyids were common in North America during the Paleocene, being represented in Europe by *Jepsenella*.

There is less anatomical evidence from the mixodectids, a fifth group of archaic placental mammals that originated in the late Cretaceous and survived into the Paleocene in Europe and North America. However, their dental and cranial anatomies are known well enough for us to have an idea of their dietary requirements. They tended to develop a rodentlike dental pattern that resembled that of the multituberculates in some ways. The mixodectids bore a pair of large, strong incisors directed forward and a cheek-tooth row composed of multicusped, low-crowned (brachydont) premolars and molars. Like the multituberculates, the mixodectids probably used this specialized dentition for crushing and opening hard seeds and nuts.

Leptictids, palaeoryctids, pantolestids, apatemyids, and mixodectids were archaic placental mammals that did not form part of a "natural," monophyletic group. However, this is not the case for all the families once included in the order Insectivora. Thus the "modern" insectivores—such as hedgehogs, shrews, and moles—belong to a monophyletic group and cannot be regarded as mere "archaic placental mammals." In fact, some of these "true" insectivores, like aquatic and terrestrial moles, bear highly specialized adaptations. The true insectivores, also known as Lypotiphla or Insectivora *sensu stricto,* were already present among the Paleocene faunas, represented by the adapisoricids, a group of extinct erinaceomorphs closely related to the hedgehog family (*erinaceomorph* means "hedgehoglike"). These early forms did not display the characteristic spiny fur of the hedgehog, but their dentition closely resembled that of today's erinaceids. Like the middle Eocene *Macrocranion,* from Messel, adapisoricids were probably small placentals about 15 cm long bearing a long tail of similar length. Their small eyes, mobile snout, and large ears suggest that they were mainly nocturnal mammals that ran on the forest floor in search of insects and fruit.

But the Cretaceous inheritance of the Paleocene was not formed only of omnivorous/insectivorous forms; small, archaic ungulates also existed in the late Cretaceous and survived the great extinction event at the end of this period. They are often known as "condylarths" (articulated condyl) because they present the kind of limb articulation related to the ungulate way of locomotion. They retained a basic, generalized design, with short limbs that ended in a five-toed foot. The distal part of the limbs was not elongated, as in modern artiodactyls and perissodactyls, and the humerus and femur were roughly of the same length as the radius and tibia, respectively. Some of the early Cenozoic condylarths, like the arctocyonids and the mesonychids, were probably also meat-eaters, and their phylogenetic position has, therefore, fluctuated as they have been classified as either primitive carnivores (creodonts) or primitive ungulates (condylarths). Most probably, they were nonspecialized omnivores that could also act as potential predators and carrion-eaters. Among them, the arctocyonids originated in the late Cretaceous and, like the multituberculates, survived to the Cretaceous–Paleocene boundary without significant losses. The arctocyonids were archaic ungulates with complete dentition, including large canines (the reason why they were often considered primitive carnivores of the order Creodonta). Their skull was long and low, with large sagittal crests and open orbits lacking a posterior bone bar. Small forms like *Protungulatum* represented the first arctocyonids in the late Cretaceous in North America. In the Paleocene, this group radiated into a variety of forms and gave rise to other groups of condylarths, such as the mesonychids, hyopsodontids, and meniscotherids.

But with the possible exception of the arctocyonids, most of these early Cenozoic mammals were, like the multituberculates, tree-dwellers living in the canopy of the rainforests that extended over most of the continents during Paleocene times. Among these arboreal mammals, the dominant ones were a kind of primatelike placental whose oldest remains, consisting of several jaws and isolated teeth of the genus *Purgatorius,* have been found in the late Cretaceous beds of Montana. Today we can easily distinguish a macaque or a chimp from any other living mammal. However, like *Protungulatum* among the first ungulates, *Purgatorius* was so close to its early "insectivorous" origins that its inclusion in the order Primates is uncertain. These primitive primatelike forms exhibited a nonspecialized anatomy, with complete dentition including large incisors, canines, premolars, and molars; an archaic skull; and extremities that more closely resembled those of a rodent than a true primate. However, a number of dental features indicate that *Purgatorius* was closer to primate origins than any other mammal of its time. Thus it began to develop enlarged central incisors, as did the Paleocene primatelike plesiadapiforms, as well as molarlike premolars and molars in which the cusps were small and low (brachydont). This dentition indicates a departure from an insectivorous diet, which involves vertical shearing by tapering pointed cusps, toward an omnivorous/frugivorous diet, in which transverse shearing and grinding dominate.

After surviving the End Cretaceous Mass Extinction, the primatelike *Purgatorius* evolved during the Paleocene into a variety of genera and species, all of them included in a single taxonomic category: the plesiadapiforms. The plesiadapiforms had a long, flattened skull with a long snout. The orbits were usually small and, in contrast to those of the younger Cenozoic primates, were not closed by a postorbital bar. Most members of the group exhibited a trend toward developing a pair of very enlarged central incisors, while the second incisors tended to be reduced or nonexistent. The limb anatomy of this group was also primitive, with a nonopposable hallux and well-developed claws—which indicates an arboreal lifestyle not very different from that of other early placentals of the same epoch. During the Paleocene and early Eocene in North America and Europe, the plesiadapiforms became one of the most successful mammalian groups, radiating into more than twenty-five genera and seventy-five species of different sizes and body weights, from some 20 g to almost 5 kg.

THE PALEOCENE MAMMALIAN RECORD IN EUROPE

In Europe, the Paleocene mammalian record is much poorer than in the rich bone-bearing beds of Montana and Wyoming in North America because during most of this epoch shallow seas covered the European territory. With the exception of some fragmentary remains recovered from the earliest Paleocene beds in Spain and Romania, the best representation of large tetrapods from Europe in these times are the dyrosaurids, a group of big crocodiles that settled the epicontinental seas and the shores of the Tethys Sea, from India to the east coast of North America, during the late Cretaceous and the Paleocene. They are included among the mesosuchians, the group of archaic crocodiles that flourished during the Mesozoic and lacked the advanced vertebral articulation of modern crocodiles, or eusuchians. After the extinction of the Cretaceous marine reptiles such as the ichthyosaurs, plesiosaurs, and mosasaurs, and before the appearance of the first cetaceans during the Eocene, these marine crocodiles were the largest vertebrate predators of the Paleocene seas.

During the middle and late Paleocene, some land emerged in the area today occupied by Belgium, northern France, and southern England. Rivers and small currents deposited marls, clays, and sandstones, carrying away the bones of the animals that lived around the marshes and the ancient lakes formed in the center of the basins. The first picture that we have from these Paleocene faunas in Europe comes from the site of Hainin, in Mons, Belgium. Somewhat younger is the late Paleocene assemblage of Cernay, France. These faunas were basically composed of the same groups as those of the late Paleocene localities of North America. However, a close inspection reveals significant differences between the mammalian associations of the two continents. Although the faunal assemblages from Cernay seem close to those of a similar age from North America, there are significant differences in the proportions of the component groups: in Europe, condylarths, followed by plesiadapiform primates and multituberculates, largely dominated the Paleocene mammalian faunas. As well, Europe lacked any relatively large ungulates such as the pantodonts, which appeared among the middle and late Paleocene faunas of North America. Finally, in contrast to the faunal exchanges that were to take place during the earliest Eocene, the late Paleocene faunas of Europe showed a high degree of endemism and isolation from North American species.

Among the multituberculates, the kogaionid, ptilodont-like *Hainina* persisted into the middle and late Paleocene in Europe. At that time, however, the European multituberculates were considerably enlarged and diversified with the entry of true ptilodonts of the genera *Liotomus* and *Neoplagiaulax*, which joined this genus. *Neoplagiaulax* and its allies bore the ptilodont dental pattern, with a large, elongated lower premolar that formed a serrated slicing blade. *Neoplagiaulax* reached a high specific diversity during the late Paleocene, with up to three species (*N. nicolai*, *N. copei*, and *N. eocaenus*).

Other late Cretaceous survivors in Europe were the marsupials, which were represented by *Peradectes*, the sole genus of this group that crossed the Mesozoic–Cenozoic boundary (figure 1.1). *Peradectes* was a didelphoid—that is, a member of the stem family of marsupials, which lay close to the origin of the modern marsupial orders. Once present on all continents, today this family has only one representative: the opossum. The opossums are active, crepuscular marsupials with climbing abilities; originally from South America, they have recently colonized

....

FIGURE 1.1 *Reconstructed life appearance of the marsupial* Peradectes
The most complete skeletons of this tiny didelphoid come from the site of Messel, Germany, and show great similarities with today's opossum species from South America. The shape of the caudal vertebrae suggests that, as in many modern opossums, *Peradectes* would have possessed a prehensile tail, one of several anatomical adaptations for life in trees. Reconstructed head and body length: 9 cm; tail, 16 cm.

eastern North America up to the forests of Virginia. *Peradectes* was a small form with a body length of about 8 cm; it displayed a long, prehensile tail of about 16 cm. This long tail suggests good climbing abilities and a lifestyle similar to that of the opossum. Like their relatives today, *Peradectes* was probably an arboreal marsupial with an omnivorous/insectivorous diet.

Among the most common mammals in Europe during the Paleocene were the primitive ungulates of the condylarth group. The archaic arctocyonids like *Prolatidens,* which were already present in Hainin, diversified in the Cernay levels into a variety of genera, such as *Arctocyon, Arctocyonides, Landenodon,* and *Mentoclaenodon. Arctocyon* was a robust arctocyonid the size of a small bear and displayed the features characteristic of the group: long skull with powerful sagittal crests and large canines (figures 1.2 and 1.3). The lower canines were even stronger than the upper ones, and when the mouth was closed, they were housed in a large space, or diasteme, between the upper canines and the premolars. Like other arctocyonids, *Arctocyon* was a plantigrade ungulate with short

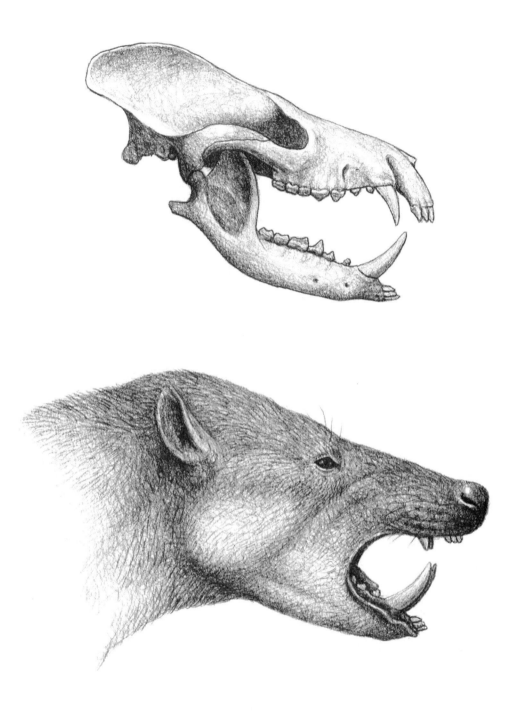

FIGURE 1.2 *Skull and reconstructed head of the arctocyonid condylarth* Arctocyon primaevus

Thanks to well-preserved cranial material from Cernay, France, we know that the skull of *Arctocyon* had an overall carnivore-like appearance, with a huge sagittal crest, a moderately long muzzle, and fearsome canines. The shape of the cheek-teeth, however, shows that meat would have made up only a small fraction of its diet, which would consist mostly of vegetable matter. Basal length of skull: 20 cm.

FIGURE 1.3 *A scene from the Paleocene site of Cernay, France*

Two arctocyonids of the species *Arctocyon primaevus* are in the foreground, while in the distance a third individual walks among the shrubs. Based on almost complete remains from Cernay, we know that these wolf-size condylarths had an unspecialized skeleton and would have vaguely resembled small bears, with their plantigrade gait and robust build. Unlike most other condylarths, *Arctocyon* had clawed feet. In the background are two flightless birds of the genus *Diatryma*. Reconstructed shoulder height of *Arctocyon:* 45 cm.

limbs, clawed feet, and a long tail. *Arctocyon* bore low-crowned bunodont molars in which the grinding surface was considerably enlarged by the addition of several accessory cusps. This dentition resembled that of present-day bears, and it is probable that *Arctocyon* had a similar omnivorous diet and lifestyle.

Arctocyonides was closely related to *Arctocyon* but smaller and slenderer. *Mentoclaenodon* was larger than *Arctocyon* and *Arctocyonides* and showed a tendency to develop long canines, like some Miocene and Pliocene big cats. However, material related to this genus is much scarcer than that of the former genera, so its exact affinities and lifestyle are still unclear.

Joining this diversified fauna of arctocyonids was *Dissacus,* a member of the peculiar family of the mesonychids (figure 1.4). The mesonychids were an unusual group of condylarths that had a specialized dentition, with tricuspid upper molars and high-crowned lower molars bearing shearing surfaces. They were probably meat- and fish-eaters and, like the arctocyonids, were once viewed as primitive carnivores. However, the limbs of the mesonychids were very different from those of the arctocyonids, having only four digits and displayed hooves supported by narrow fissured terminal phalanxes. The only European mesonychid, *Dissacus,* was the size of a wolf and during the late Paleocene attained a Holarctic distribution, from North America to Europe and Asia.

The remaining condylarths of the European Paleocene were mainly specialized vegetarian mammals that distantly fit the modern conception of ungulates. These were the hyopsodontids and the meniscotherids. The hyopsodontids were archaic condylarths that had an ubiquitous distribution throughout Eurasia and North America. All of them were small ungulates, their size ranging from that of a squirrel to that of a weasel. Although much more herbivorous in their diet than the arctocyonids, and lacking their powerful canines, the hyopsodontids still had a generalized dentition, with a full set of incisors, canines, premolars, and molars. During the Paleocene in Europe, they reached a high diversity level, starting with *Louisina* and *Monshyus* in Hainin and following in the Cernayssian beds with *Tricuspiodon, Paratricuspiodon,* and *Paschatherium.*

Paschatherium was a small condylarth with an insectivore-like dentition that closely resembled that of the primitive erinaceomorph insectivores such as *Adapisorex* (until recently, some authors included this genus among the insectivores, rather than among the condylarths). *Paschatherium*'s tarsal and limb morphology indicates that it was an arboreal form well adapted to climbing and running in trees. Moreover, *Paschatherium* is interesting because of its possible affinities with the group of endemic ungulates of Africa, or Tethytheria. More specifically, its tarsal anatomy suggests a straight relationship with today's hyracoids (the hyraxes of the African savannas). Therefore, *Paschatherium* could be seen as the remote ancestor of such different groups as the hyraxes, elephants, sea cows, and aardvarks, as well as extinct groups that radiated in Africa during the Eocene.

A second family of herbivorous condylarths was the meniscotherids, a group that displayed more specialized characteristics than the generalist hyopsodonts. The members of this family that settled in Europe during the Paleocene, like *Orthaspidotherium* and *Pleuraspidotherium,* showed some of the characteristics that were present in such ungulates as the artiodactyls (figures 1.5 and 1.6). Thus the meniscotherids' premolars resembled their own molars, the dentition thus forming a unique battery of similar cheek-teeth (we say, therefore, that the premolars were "molarized"). Also, their cheek-teeth developed recurved, crescent-shaped shearing ridges like those seen in present-day rumi-

FIGURE 1.4 *Reconstructed life appearance of* Dissacus europaeus

The fossil record of this species is rather fragmentary, so this reconstruction is partly based on better-known, closely related mesonychyd genera from Asia and North America, like *Synoplotherium* and *Pachyaena*. Remains of *Dissacus* from Cernay, including a mandible, a complete radius, and fragments of the humerus, indicate that the European species was about the size of a modern jackal. A study of the morphology of the limb bones suggests that the animal was digitigrade and more cursorial than usually thought of members of this genus. Reconstructed shoulder height: 33 cm.

nants (what is called a selenodont dentition, after *selene*, the Greek word for "moon"). This indicates that *Pleuraspidotherium* and its relatives were probably browsers, maintaining a diet based on leaves and even harder vegetation. *Pleuraspidotherium*'s skeleton, however, was still fully condy-

larthian in shape, lacking the cursorial adaptations present in the artiodactyls and other advanced ungulates.

Besides the condylarths, the group second in importance in the Paleocene rainforests of Europe was the primatelike plesiadapiforms,
which attained great diversity. The plesiadapiforms were mainly insectivorous, bearing a pair of enlarged, lanceolate lower incisors. The most
primitive and smallest plesiadapiforms were members of the microsy-
opid family, like *Berruvius. Berruvius* was a typical member of this family,
weighing only 20 g, the size of a shrew. However, the most common and
best-known plesiadapiform was the plesiadapid *Plesiadapis,* from which
several skulls and partial skeletons have been recovered in the Cernayssian levels in France (figure 1.7). Like the ptilodont multituberculates
and the mixodectids, *Plesiadapis* developed a rodentlike skull design,
with a long snout ending in a pair of long, chisel-like incisors in each

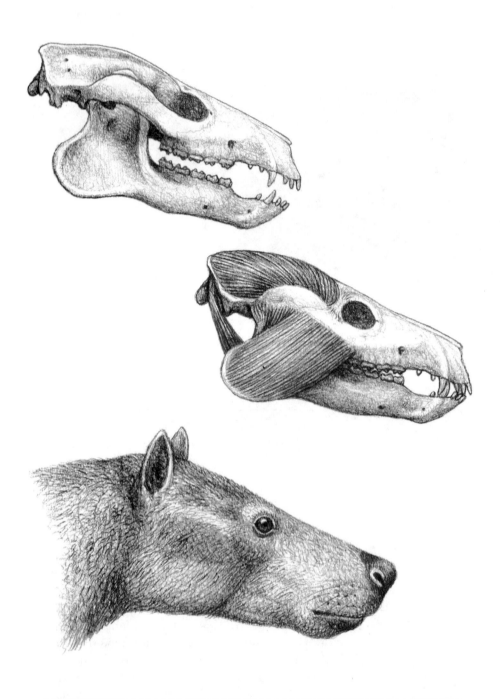

FIGURE 1.5 *Skull, masticatory muscles, and reconstructed head of the meniscotherid* Pleuraspidotherium aumonieri *from Cernay*

In contrast with the omnivorous *Arctocyon*, *Pleuraspidotherium* displayed clear adaptations for a vegetarian diet in both dentition and muscle insertions. The great development of the masseter muscle and the relatively weak temporalis indicate clear emphasis on the chewing function of the cheek-teeth. Total length of skull: 12 cm.

jaw, separated from the cheek-teeth by a long diasteme. *Plesiadapis* was once thought to be an arboreal primate because of its long tail and long, laterally compressed claws, but the robustness of the skeleton and the limb proportions suggest more terrestrial habits. In fact, given its wide spatial distribution, *Plesiadapis* is now thought to have been an arboreal quadruped also able to feed and move on the ground.

A second plesiadapid genus, *Chiromyoides,* was smaller than *Plesiadapis* (about 300 g, against the 1 to 2 kg of the latter) and had a shorter snout and deeper mandible. Although sharing with *Plesiadapis* a pair of strong incisors, *Chiromyoides* probably had a more specialized diet based on seeds.

Another group of plesiadapiform primates, the North American carpolestids, paralleled the rodentlike pattern of both the plesiadapids and the ptilodont multituberculates. The carpolestids not only developed strong, chisel-like incisors but also modified their last lower premolar into a large serrated slicing blade like that of *Ptilodus.*

In Europe, the plesiadapiform *Saxonella* evolved in a similar way, and for years it was thought to be a true carpolestid. However, further analysis has shown that the enlarged premolar was not the fourth lower premolar but the third one, indicating that this "ptilodont-like" dentition evolved independently in the course of the Paleocene in at least three completely different evolutionary lineages!

In addition to the condylarths and the plesiadapiforms, other common members of European Paleocene biotas were the archaic placental mammals, grouped in the families Leptictidae, Pantolestidae, Apatemyidae, Mixodectidae, and Palaeoryctidae. The most diversified among them were the generalist leptictids, with up to three genera (*Paleictops, Adunator,* and *Diaphyodectes*). The other groups of archaic placental mammals were less diversified and were usually represented by a single genus: the otterlike pantolestids (*Pagomomus*), the rodentlike mixodectids (*Remiculus*), the insect-eater apatemyids (*Jepsenella*), and the palaeoryctids (*Aboletylestes*). Like the leptictids, the "true insectivore," hedgehog-related adapisoricids attained high diversity levels during the Paleocene, with genera like *Adapisoriculus* and *Leptacodon.*

THE NONMAMMALIAN PREDATORS

The arctocyonids were among the largest mammals of the European Paleocene, reaching the size of a small bear and having a mainly omnivorous diet. The mesonychid *Dissacus* probably had more carnivorous habits but was certainly smaller (the size of a coyote), and its diet probably consisted mainly of fish. Does this mean that the herbivorous or omnivorous mammals like *Paschatherium, Pleuraspidotherium,* and *Orthaspidotherium* that settled in the canopy forests and marshes of the Euro-

FIGURE 1.6 *Reconstructed life appearance of* Pleuraspidotherium aumonieri
Partial remains from Cernay indicate that this animal had a generalized skeleton. Like *Arctocyon,* it walked with rather abducted limbs and was plantigrade, but it was considerably smaller and more slenderly built. Reconstructed shoulder height: 23 cm.

pean Paleocene lived a pleasant existence without large predators? Not at all. In fact, the presence of large, dangerous, nonmammalian predators continually stressed the Paleocene dwellers of these canopy forests and marshes. These predators included the gigantic birds of the diatryma group and the eusuchian crocodiles, which during these early Cenozoic times can be regarded as the true heirs of the carnosaurian and coelurosaurian dinosaurs that filled the large-predator guild in the late Cretaceous.

The diatrymas were human-size—up to 2 m tall—ground-running birds that inhabited the terrestrial ecosystems of Europe and North America in the Paleocene and the early to middle Eocene (figure 1.3). They were flightless birds, without functional wings, and had short but strong legs. *Diatryma*'s most outstanding feature, however, was its large, powerful beak, which was supported by a large skull and a robust neck. Once considered as close relatives of the rails (or gruiformes), diatrymas, according to recent analysis, appear to be related to waterfowls (or anseriformes). During the Paleocene in Europe, the ostrich-size *Gastornis* represented the diatrymiformes; like the American *Diatryma,* it was probably a meat- and carrion-eater.

Besides the large diatrymas, a large variety of crocodiles—mainly terrestrial and amphibious eusuchian crocodiles—populated the marshes of the Paleocene rainforests. The eusuchians appeared during the Cretaceous and were characterized by the advanced vertebral articulation that we see in today's crocodiles and alligators. The members of this group have asymmetrical vertebrae, with a concave anterior part and a convex posterior one. The anterior rounded face of each vertebra fits the convex side of the preceding one, and this structure allows the characteristic flexibility and lateral torsion of modern crocodiles. After the extinction of the last dinosaurs, the eusuchians flourished and diversified throughout the Paleocene and Eocene terrestrial ecosystems. Among the largest was *Asiatosuchus,* a true crocodile of the Nile croco-

FIGURE 1.7 *Reconstructed skeleton of the primitive primate* Plesiadapis tricuspidens

Well-preserved skeletal remains from Cernay reveal that this animal was rather similar in structure to a squirrel, with comparable climbing adaptations. The shape of the articulation of the elbow shows that the forearm could not be completely extended, giving the animal a crouched stance typical of quadrupedal climbers. Like squirrels, but unlike modern primates, *Plesiadapis* had claws instead of nails. Reconstructed shoulder height: 25 cm.

dile group; it was about 4 m long and had a powerful skull 70 cm long. Like its present-day relatives, it used its strong tail as a propelling structure in the water.

Coexisting with *Asiatosuchus,* the alligator group reached smaller dimensions but attained a higher diversity. The most common form was *Diplocynodon,* a caimanlike crocodile about 1.5 m in length. Its short snout indicates that it was mainly a flesh-eater.

Similar in size was *Pristichampsus,* another alligator-like crocodile that developed flattened, serrated teeth like those of some carnosaurian dinosaurs. Unlike *Diplocynodon, Pristichampsus* bore a distinctly elongated snout armed with blade-shaped teeth; this dentition indicates a diet based mainly on fish.

Joining *Asiatosuchus, Allognathosuchus* was a small alligator about 1 m long that showed a high degree of regional specialization in its teeth. Thus while a part of its dentition indicates an adaptation to catching and breaking up prey at the moment of capture, the presence of rounded teeth suggests that it also supplemented its diet with mollusks and other hard organisms. All these crocodile forms succeeded during the Paleocene and Eocene in Europe and North America, and some of them, like *Diplocynodon,* even lasted into North America's Miocene.

The high diversification of the crocodile fauna throughout the Paleocene and Eocene represents a significant ecological datum, since crocodiles do not tolerate temperatures below 10 to 15°C (exceptionally, they could survive in temperatures of about 5 or 6°C). Their existence in Europe indicates that during the first part of the Cenozoic the average temperature of the coldest month never fell below these values and that these mild conditions persisted at least until the middle Miocene.

THE END OF THE PALEOCENE

In North America, but also in Asia, the late Paleocene faunas were considerably enriched with several orders of new mammals. Strange, noncondylarthian ungulates reached sizes unknown for a mammal at that time. These were the pantodonts, a group of bizarre herbivores that attained the size of a rhinoceros. They bore strong, tusklike upper canines, which in some forms, like *Titanoides,* were long and resembled those of the saber-toothed cats. However, despite the pantodonts' large dimensions, several aspects of their anatomy reveal that they were rather archaic placentals, with W-shaped ridges in the upper molars (as in most insectivores), heavily built bodies, and short, stout limbs with five digits. After their dispersal from Asia to North America during the middle Paleocene, they became in the late Paleocene a common member of the terrestrial faunas of both continents. One genus in particular, *Coryphodon,* attained at that time a wide distribution in the northern continents (figure 2.2).

Another group of North American subungulates, the taeniodonts, developed large, massive herbivorous forms like *Psittacotherium* during the late Paleocene. In this group, the canines reached great dimensions, while the incisors and the premolars acquired a caninelike shape. At the same time, the molars became high-crowned (a phenomenon known as "hypsodonty"), becoming empty prisms without cusps. This specialized dentition was supported by powerful mandibles, which, in turn, were supported by a strong musculature. The anterior limbs became more and more robust, adapted to digging. It is possible that the hypsodont cheek-teeth of the taeniodonts were a response to a diet based on hard vegetation or perhaps a consequence of the digging process, which always involves the ingestion of a certain quantity of abrasive sand and soil particles.

As with the multituberculates and the plesiadapiform primates, other late Paleocene herbivores developed a rodentlike design, with long, high-crowned incisors. These were the tillodonts, a group of large, massive mammals of unknown affinities that, according to some researchers, were related to the condylarths. The tillodonts' limbs were short and ended in plantigrade, five-toed feet. Their incisors were chisel-like and grew continuously, as do those of rodents and rabbits. However, their cheek-teeth were archaic, formed of low-crowned (brachydont), cusped molars (or bunodont molars, as opposed to those bearing crests, which are called lophodont). Primitive forms like *Esthonyx* entered North America from Asia during the late Paleocene, succeeding and diversifying until the middle Eocene. From North America, they made a short incursion into Europe during the early Eocene.

But at the end of the Paleocene, after several rodentlike "experiments," the true rodents finally entered onto the scene. They migrated from Central Asia, where they most probably originated, to North America and later, during the early Eocene, to Europe. The first known rodents (*Arctoparamys*) belonged to the family Ischyromyidae and were squirrel-like arboreal forms with basic dental and limb anatomies. However, they already displayed the characteristics typical of the order, with a pair of ever-growing incisors in each jaw, separated from the cheek row by a long diastema without teeth. After their appearance during the late Paleocene, they soon spread and became the dominant group of small mammals, a position they have retained throughout the Cenozoic. Soon after the appearance of the true rodents, the multituberculates suddenly decreased in diversity, becoming much rarer until their extinction during the Oligocene.

Rodents were not the only order of modern mammals to appear at the end of the Paleocene; the true carnivores appeared at the same time in North America. They were the miacids, a group of small, archaic carnivores that probably preyed on rodents and other arboreal dwellers of the late Paleocene forests.

The faunas of North America and Asia changed considerably over the course of the Paleocene. Other than small, insectivorous forms dwelling in the canopies of the rainforests, a number of large-bodied mammals already existed at the end of the epoch. Most significantly, the first representatives of the orders of mammals that were going to dominate the terrestrial faunas in the next epochs began to appear. The picture was different in rather isolated Europe, which at the end of the Paleocene still retained a fauna composed basically of arboreal mammals (primates, multituberculates, and some condylarths), insectivores, some small ground mammals, and incipient browsers (hoofed condylarths). But the conditions were going to change for Europe also.

CHAPTER 2
The Eocene: Reaching the Climax

AT THE END OF THE PALEOCENE, APPROXIMATELY 55.5 MILlion years ago, there was a sudden, short-term warming known as the Latest Paleocene Thermal Maximum. Over a period of tens of thousands of years or less, the temperature of all the oceans increased by around 4°C. This was the highest warming during the entire Cenozoic, reaching global mean temperatures of around 20°C. There is some evidence that the Latest Paleocene Thermal Maximum resulted from a sudden increase in atmospheric CO_2. Intense volcanic activity developed at the Paleocene–Eocene boundary, associated with the rifting process in the North Atlantic and the opening of the Norwegian-Greenland Sea. Massive eruptions took place, extruding about 2 million cubic km of basalts (Roberts et al. 1984) and injecting into the atmosphere considerable quantities of greenhouse gases, such as CO_2 and CH_4. According to some analyses, atmospheric CO_2 during the early Eocene may have been eight times its present concentration. In any case, the sudden increase of temperature led to the extinction of large numbers of protists and small planktonic organisms that were adapted to the cooler conditions of the Paleocene. The high temperatures and increasing humidity favored the extension of tropical rainforests over the middle and higher latitudes, as far north as Ellesmere Island, now in the Canadian arctic north. There, an abundant fauna—including crocodiles, monitor lizards, primates, rodents, multituberculates, early perissodactyls, and the pantodont *Coryphodon*—and a flora composed of tropical elements indicates the extension of the forests as far north as 78 degrees north latitude. These early and middle Eocene forests were more diverse and included more tropical elements than those of the late Paleocene (up to 92% of tropical species, against 42% of potential temperate elements; Collinson and Hooker 1987). They were comparable to the paratropical forests of eastern Asia, formed from dense, stratified arboreal vegetation that included lianas, shrubs, and herbaceous climbers, in a context of high humidity.

The global oceanic level at the beginning of the Eocene was high, and extensive areas of Eurasia were still under the sea. In this context, Europe consisted of a number of emerged islands forming a kind of archipelago. A central European island consisted of parts of present-day England, France, and Germany, although it was placed in a much more southerly position, approximately at the present latitude of Naples. Important early Eocene fossiliferous localities like Dormaal in Belgium, Avenay in France, and the Bembridge beds in England were deposited on this emerged land. An Iberian plate including Portugal and central Spain formed a second large island farther to the south. The significant fossiliferous site of Silveirinha, close to Lisbon, was located on this island. To the east, the growing Mediterranean opened into a wide sea, since the landmasses of Turkey, Iraq, and Iran were still below sea level. To the east of the Urals, the Turgai Strait still connected the warm waters of the Tethys Sea with the Polar Sea.

PALEOCENE SURVIVORS IN THE EARLY EOCENE

Despite its dramatic effects on the ocean, the Latest Paleocene Thermal Maximum seems not to have deeply affected the early Eocene terrestrial faunas. The only direct effect that can be identified is the extension to the north of taxa so far restricted to the lower and middle latitudes, such as multituberculates, primates, pantodonts, crocodiles, and monitor lizards. This warming of the higher latitudes, with the subsequent advance of the forests to the north, probably had another indirect effect, favoring the interchange across the land bridges that at that time still connected Europe and North America. In this way, the early Eocene records a sudden entry into Europe of several of the mammalian taxa that were present during the late Paleocene in North America, such as tillodonts, pantodonts, miacid carnivores, and ischyromyid rodents. These newcomers joined a number of relics of late Paleocene origin that survived into the early Eocene in Europe. These Paleocene remnants, although greatly reduced in number, consisted basically of multituberculates, condylarths, and plesiadapiforms.

The first rodents to enter Europe probably had immediate effects on the autochthonous multituberculate faunas, leading to the extinction of the archaic kogaionids of the genus *Hainina*. However, although rare, the ptilodont multituberculates of the genus *Neoplagiaulax* survived for some time into the early Eocene. Among the marsupials, the situation was quite different, since the group of opossum-like didelphoids not only survived but even significantly increased in diversity. Thus besides the Paleocene *Peradectes* (which was, in fact, a late Cretaceous survivor!), two other opossum-like genera, *Peratherium* and *Amphiperatherium*, evolved during the early Eocene. From some well-preserved specimens of *Amphiperatherium* from the middle Eocene of Messel, we know that these newcomers were somewhat larger than *Peradectes* (with a body length of around 15 cm) but had relatively shorter tails (about 17 cm) (figure 2.1). This indicates that they probably had a less arboreal existence, living most of the time on the ground. The discovery of some early Eocene didelphoid relatives in northern Africa has raised the idea that these marsupials entered Europe directly from South America, across a South American–African–Iberian corridor. The existence of such a migratory route makes some sense in view of a number of middle Eocene findings that are discussed later. Other groups of small mammals that were not affected by the entry of the new Eocene immigrants were those included among the insectivores, either the "true" insectivores, like the erinaceomorph adapisoricids, or the archaic placentals, like the apatemyids. Particularly, the apatemyids appear as a rather common group in the Eocene forests of Europe, being represented by a number of genera and species, such as the early Eocene *Apatemys*.

Among the most significant groups of Paleocene survivors were the primatelike plesiadapiforms, such as the relatively large *Plesiadapis* (which reached weights of up to 5 kg). The plesiadapids even increased their diversity with the development of another large form, *Platychoerops*, of about 3 kg.

A second group of plesiadapiforms surviving into the early Eocene was the paromomyids, represented in Europe by the genera *Phenacolemur* and *Arcius*. Both were small to medium-size plesiadapiforms, *Phenacolemur* weighing between 150 and 500 g and *Arcius* weighing about 160 g. They bore a pair of enlarged lower incisors that operated against another pair of lobate incisors in the upper jaw. A long diasteme, resulting from the reduction and loss of the canines and premolars, separated the incisors from the molars. The molars showed a flat, broad surface, and the last molar was characteristically elongated, suggesting a herbivorous diet based on fruits.

The early Eocene still retained a diversified fauna of condylarths, formed basically of hyopsodontids and, to a lesser degree, arctocyonids. Among the arctocyonids, *Landenodon* persisted from the late Paleocene. Similar in many respects to the arctocyonids were the paroxyclaenids, a group of moderately small mammals with large, grooved canines and low- to moderately high-crowned molars. Because of its archaic dental and limb anatomies, this group had been thought to be related to the arctocyonids. However, Russell and Godinot (1988) contested the arctocyonid affinities of the paroxyclaenids, proposing instead a close relationship to the pantolestid "insectivores." From the early Eocene *Merialus*, in the French locality of Palette, during the middle and late

Eocene the pantolestids diversified into a number of genera, such as *Paroxyclaenus* from Belgium, *Spaniella* from Spain, and *Kopiodon* from Germany. According to the specimens of *Kopiodon* preserved in Messel, these pantolestids were arboreal mammals, about 1 m long, that still retained their clavicles as a characteristic feature. (Animals running on the ground usually lose their clavicles, since the limbs do not have to extend to the sides and there is no need to stabilize the shoulder girdle.) Moreover, two bones of the forearm, the radius and ulna, were completely separated, allowing the rotation of the hand (the pronation-supination movement). All these features, as well as the presence of large, laterally flattened bony claws at the ends of the fingers, confirm the climbing abilities of these mammals and an arboreal lifestyle. According to their tooth morphology, the paroxyclaenids were mainly fruit-eaters.

Nevertheless, the most common condylarths of the early Eocene were the hyopsodontids, represented by *Microhyus* (closely related to the Paleocene *Louisina* and *Monshyus*), *Paschatherium*, and *Hyopsodus*. In fact, the small hyopsodontid *Paschatherium* was the dominant element in the early Eocene assemblage of Dormaal, Belgium. Closely related to *Paschatherium* was *Hyopsodus*, which, unlike the former, was not so well suited for an arboreal mode of life. *Hyopsodus* was probably a semiterrestrial condylarth with a tarsal anatomy suggesting fossorial capabilities. The two ramus of the lower jaw were fused at the symphisal region and bore long procumbent, forward-directed incisors. In the upper jaw, the incisors were large and vertically implanted. This is a quite specialized dentition for such a small mammal, resembling the one borne by the ancestors of the African ungulates, and it reinforces the idea of affinity between the hyopsodontids, such as *Paschatherium* and *Hyopsodus,* and the ungulate tethytheres of Africa, such as the hyraxes and the proboscideans. In any case, the European hyopsodontids had to face the competition of the more advanced condylarths of the genus *Phenacodus*, which in the early Eocene entered from North America. Unlike the archaic condylarthians, the phenacodontids were better suited for cursorial locomotion, approaching in this way the modern order of the perissodactyls. *Phenacodus* was the size of a sheep, about 1.5 m in length

....

FIGURE 2.1 *A scene from the middle Eocene site of Messel, Germany, with the early carnivore* Parodectes feisti *and the marsupial* Amphiperatherium *climbing down a liana*

A fortuitous encounter like this one between a miacid carnivore and an opossum could possibly have had a sad ending for the latter, since the marten-size *Parodectes* was an agile climber. It did not have a prehensile tail, as did the marsupial, but was probably much faster, and its versatile skeleton with flexible limbs was well adapted to grasping branches and jumping between tree limbs. Total length of *Parodectes:* 55 cm.

and 60 cm at the withers (plate 1). It had a long tail and limbs with five hoofed digits that could rotate sideways. This group might be related to the first odd-toed ungulates, or perissodactyls.

This diversified fauna of condylarths declined over the course of the early Eocene, and by the end of this period only the advanced phenacodontids survived. Other late Paleocene remnants declined in a similar way, so that at the end of the Eocene no multituberculate existed in Europe, while the plesiadapiforms became truly rare. The appearance and spread of more specialized forms like the perissodactyls and other representatives of the modern orders of mammals probably caused this decline.

THE NORTH AMERICAN CONNECTION

Despite the opening of the Greenland-Norwegian Sea, Europe and North America were still connected during most of the early and middle Eocene across two main land bridges. The northern one, the De Geer Corridor, connected Norway to Greenland and, from there, to Ellesmere and Baffin Islands in Canada. To the south, the Thule Bridge connected Scotland and the central European territory to the Faroe Islands, Iceland, Greenland, and, from there, again to Ellesmere and Baffin Islands. Under the humid and warm conditions of the early Eocene, these corridors must have been effective, since the European fossil record shows a massive entry of American elements, such as pantodonts, tillodonts, carnivores, creodonts, rodents, and dermopterans.

Among the most impressive mammals to invade Europe from North America in the early Eocene were the tillodonts of the genus *Esthonyx* and the large pantodonts of the genus *Coryphodon*. *Esthonyx* was a primitive tillodont with enlarged lateral incisors but lacking the diasteme that was present in the advanced members of the group. *Coryphodon* was the size of a hippo and, like it, probably had an amphibious lifestyle (figure 2.2). It was a massive, heavily built pantodont with short feet and lacked the characteristic long tail of the condylarths. The members of this genus displayed strong upper canines separated from the cheek-teeth row by a short diasteme. The cheek-teeth series was not reduced, with a complete set of molars and premolars that superficially resembled those of today's tapirs. Like them, *Coryphodon* was probably a semiaquatic browser inhabiting the land close to marshes and rivers.

At the same time that the large-herbivore guild grew in such an extraordinary way, the carnivore guild was also considerably enriched by the entry of the first true carnivores of the family Miacidae and a group of archaic meat-eaters included in the separate order Creodonta.

The creodonts were a group of extinct meat-eaters that dominated the predator guild from the late Paleocene to the Oligocene, when they

FIGURE 2.2 *Reconstructed life appearance of* Coryphodon
Although widespread during the Eocene in Europe, this genus is best known thanks to well-preserved skeletons from various sites in North America. That material shows the skeleton of *Coryphodon* to be a mixture of traits reminiscent of those of different kinds of animals. The trunk vertebrae have surprisingly weak neural spines for such a big animal, suggesting a partly amphibious lifestyle, like that of modern hippos. The long bones of the limbs resemble in structure those of heavy perissodactyls like rhinos and tapirs, while the feet, retaining all five digits, are like those of modern elephants in structure. In side view, the head vaguely resembled that of the Paleocene arctocyonids, with huge canines, although this animal was not an omnivore like the latter, but a specialized vegetarian. Reconstructed shoulder height: 1.2 m.

were finally replaced by more specialized true carnivores. They probably evolved from a late Cretaceous or Paleocene palaeoryctid "insectivore" that transformed the central part of its dentition into a shearing blade capable of cutting and processing meat. The upper and lower bladelike teeth modified in this way are, therefore, called "carnassials." In today's carnivores, the last upper premolar and the first lower molar invariably have this function, but in the creodonts either the first upper molar and the second lower molar, or the second upper molar and the third lower molar, played this role. Therefore, the creodonts' carnassials were in a more rearward position than those of the true carnivores. There were also other differences—for instance, in the skull, which used to be long and massive. The jaw joint was almost spherical, enabling a greater mobility of the mandible. However, the creodonts were in general less well suited for active hunting than modern carnivores, since most of them were plantigrades and had less carpal and tarsal mobility.

During the early Eocene, the diversity of this group suddenly increased in Europe with the entry of *Oxyaena, Paleonictis, Prolimnocyon,* and *Proviverra*. *Oxyaena* was a primitive member of the American family of the oxyaenids. It was a large predator for its time, about 1 m long. The first upper molar and the second lower molar formed the carnassials, which showed the development of slicing blades (although the puncture-crushing function was still dominant). Much heavier jaws and blunt cusped teeth, probably used for bone crushing, characterized another oxyaenid, *Paleonictis*. Both *Oxyaena* and *Paleonictis* were robust plantigrade forms supported by short, heavy limbs. *Proviverra*, *Arfia*, and *Prolimnocyon* are also classified as creodonts, although they are placed in a separate group, the hyaenodonts, whose more carnivorous dentition and carnassials, formed in this case by the second upper molar and the third lower molar, made them different from the oxyaenids.

The hyaenodonts were better suited for active hunting than the oxyaenids, developing in some cases a semidigitigrade or fully digitigrade locomotion, as is found in most modern carnivores. However, the early Eocene forms like *Proviverra* had not yet developed good running abilities (plate 1). They were small hyaenodonts of about 20 cm that probably fed on small mammals, birds, and even insects. *Prolimnocyon* was also a small, generalist hyaenodont that probably passed most of its time hunting in the undergrowth or climbing in trees. It had a long snout that it used to scratch in the ground for smaller vertebrates and insects. In contrast, *Arfia* had larger dimensions (the size of a fox) and possessed teeth with better developed shearing surfaces, which it used for slicing meat. It was a cursorial form, with elongated limbs and digitigrade feet. Despite its running abilities, there was some capacity for hind-foot reversal, suggesting that it may also have been a good climber.

THE BEGINNING OF THE MODERN ORDERS OF MAMMALS

Together with the creodonts, the first miacid carnivores entered Europe during the early Eocene. These first true carnivores still had a nonspecialized puncture-crushing dentition, so their origin cannot be related to that of the palaeoryctids and creodonts. Unlike the creodonts, the miacids used the fourth upper premolar as the upper carnassial, while the first lower molar played the function of the lower carnassial. The anterior part of the tooth (or the trigonid) evolved into a blade for slicing, while the posterior part (known as the talonid) retained its puncture-crushing function, as did the postcarnassial molars. In their turn, the premolars of the precarnassial region shrank, being used mainly to hold prey. This division of the dentition into three functional regions (precarnassial, carnassial, and postcarnassial) enabled the carnivores to attain great adaptive flexibility and played a fundamental role in the success of this order of mammals.

Like the miacids, the ischyromyids, the family of primitive rodents that appeared during the late Paleocene in North America, invaded Europe for the first time in the early Eocene. They diversified there into a number of genera, such as *Plesiarctomys, Pseudoparamys,* and *Microparamys. Pseudoparamys* and, especially, *Plesiarctomys* were quite large ischyromyids, the latter about the size of a marmot. In contrast, *Microparamys* included small, mouse-size species like *M. chandoni* and *M. russelli.* This genus is of high evolutionary significance because some of its species fit as ancestors of the successful later Eocene and Oligocene families such as the glirids and the theridomyids.

A second group of archaic rodents, the ailuravids, joined the ischyromyids in the early Eocene. A first species, *Meldimys louisi,* invaded Europe from North America at this time, being replaced shortly afterward by *Ailuravus.* Both were characterized by their archaic dentition, which was even less complex than that of the coeval ischyromyids. Thanks to the well-preserved specimens of *Ailuravus* recovered from the middle Eocene at Messel, we know that these were large rodents of about 40 cm that bore a robust, 10-cm skull and a long, 60-cm tail. A dense fur covered the back, while the long tail was also bushy. Sharp claws, a common feature among several tree climbers, armored the toes. *Ailuravus*'s dentition was low crowned and displayed heavily crested, crenulated enamel, as in some modern flying squirrels that base their diet on leaves. *Ailuravus* was probably an arboreal leaf-eater whose lifestyle likely resembled that of extant giant squirrels of eastern Asia (*Ratufa*). The gut contents from the best-preserved specimens of Messel, which contained fragments of *Polyspora* leaves, a member of the tea family found today in Southeast Asia, confirms these dietary requirements of *Ailuravus.*

The ischyromyid and ailuravid rodents, as well as the miacid carnivores, were among the oldest representatives of the modern orders of mammals to appear in Europe during the early Eocene. However, they were not the only ones, since the "modernization" of the mammalian communities at this time went even further, and groups such as the first true primates, bats (Chiroptera), flying lemurs (Dermoptera), and odd-toed (Perissodactyla) and even-toed (Artiodactyla) ungulates entered onto the scene, in both Europe and North America.

One of the most spectacular stories is that of the bats, or chiropterans. Absent from the Paleocene, they suddenly appeared during the early Eocene in North America (Wyoming) and Europe (Paris Basin). What is most surprising is that these first chiropterans were completely formed bats that differed from their modern relatives only in some anatomical details. Although one could imagine a number of transient stages linking the patagium of a flying squirrel to the digit-supported wing of a bat, these are completely theoretical reconstructions that have never been recorded as fossils: the first chiropterans bore all the characteristics recognized in present-day bats, having a similar predatory behavior and lifestyle. They appeared abruptly during the early Eocene, represented by a variety of genera, such as *Icaronycteris, Archaeonycteris,* and *Palaeochiropteryx.*

Icaronycteris had some primitive features, such as the retention of a claw in its index finger (a characteristic that is absent from today's insectivorous bats, or microchiroptera). It had a straight skull with small orbits and a long tail with free vertebrae. Like *Icaronycteris, Archaeonycteris* retained some primitive features, such as the presence of a claw in the index finger. With a wingspan of about 35 cm, it was a relatively large form that closely resembled today's large mouse-eared bats (the vespertillionids) and horseshoe bats (the rhinolophids), all characterized by their nonspecialized flight in open spaces.

In contrast, *Palaeochiropteryx* was a small, specialized bat with relatively large wings of between 25 and 30 cm (figures 2.3 and 2.4). Having

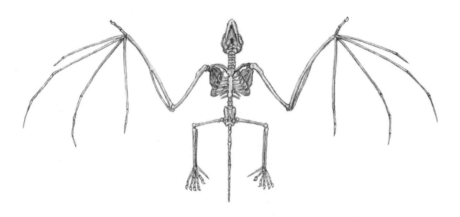

FIGURE 2.3 *Skeleton of the early chiropteran* Palaeochiropteryx tupaiodon
Complete and articulated specimens from Messel provide an accurate picture of the body proportions and skeletal morphology of *Paleochiropteryx*. While several features of the skull and skeleton are clearly primitive, including the shape of the elbow articulation and the proportions of the forearm, the general proportions of the wings already resemble those of such modern bat families as the hipposiderids in being remarkably short and broad (see figure 4.6). Such wing proportions are early adaptations for slow but maneuverable flight amid the undergrowth and branches of the forest. Reconstructed wingspan: 25 cm.

FIGURE 2.4 *Reconstructed life appearance of the bat* Palaeochiropteryx tupaiodon *in flight*
Modern bats often drink on the wing, skimming over the surface of lakes and rivers, and it is possible that the same habit was one of the reasons for the preservation of so many bats in the site of Messel. It has been hypothesized that the paleolake at Messel would have occasionally emitted toxic gases, and any bat flying near the surface would have been killed. Also, the wing proportions of *Paleochiropteryx* show that it was a low-altitude flier, hunting for insects near ground level and near the lake surface. This would help explain why it is the most commonly found bat in the site, while other species with wing proportions indicating high-altitude flight are remarkably less abundant. Reconstructed wingspan: 25 cm.

relatively broad wings reduces wing-loading and permits slow flight. Consequently, these animals had a high capacity to avoid the obstacles that could appear during flight close to the ground or in closed spaces. This kind of extreme specialization in bats is seen among modern leaf-nosed bats, or hipposiderids, which have an agile flight path low to the water or ground. Similarly, *Palaeochiropteryx,* with its relatively large wings and reduced wing-loading, was probably adapted to low flight through the dense foliage of Eocene forests and just above the waters of small lakes and marshes.

Another group of "flying" mammals to enter Europe was the first dermopterans of the family Plagiomenidae. The dermopterans, or flying lemurs (*Cynocephalus*, today's colugo), belong to a specialized order of gliding mammals that today inhabits the dense forests of Southeast Asia. They were relatively small mammals (1 to 2 kg) that developed large skin folds connecting their otherwise elongated arms and limbs. These folds formed a large patagium that enabled them to glide from tree to tree over distances of more than 100 m. Their diet was composed mainly of fruits and leaves, which the dermopterans harvested with specialized, comblike incisors. The first dermopterans of the genus *Plagiomene* appeared during the late Paleocene in North America and entered Europe during the early Eocene, together with the first carnivores and rodents. *Plagiomene* already bore spatulate and forked incisors closely resembling those of today's *Cynocephalus*. Anatomical and molecular evidence indicates that the dermopterans probably evolved from some archaic Paleocene primatelike placentals, and, in fact, recent data suggest that they are closely related to the plesiadapiforms, notably the mostly primitive microsyopids.

But the most successful Eocene arboreal mammals were the true primates of the families Adapidae and Omomyidae, which replaced the plesiadapiforms. These true primates, or euprimates, were quite different from the archaic plesiadapiforms, having features common to any modern member of the order. Many of the differences relate to the true primates' full adaptation to an arboreal lifestyle. Unlike the plesiadapiforms, the true primates had shorter snouts and a postorbital bony bar forming a ring around the orbit—which indicates the lesser importance of the sense of smell and the higher reliance on vision. The claws of the plesiadapiforms were replaced by nails, and the hallux became opposable—which would allow for increased leaping abilities. The true primates' dentition was also different, having much reduced lower incisors and dominant, increasingly enlarged canines instead of the rodentlike design of most of the plesiadapiforms. Moreover, adapids and omomyids may have been related to some groups of today's primates, such as the lemuriforms (in the case of the adapids) and the tarsiforms (in the case of the omomyids).

The adapids were the most widely diversified and successful group of primates in the North American and European Eocene, when they split into several genera and species. The skull of the adapids had a relatively long and broad snout and, in this way, resembled that of such modern prosimians as the lemurs of Madagascar. Like today's lemurs, some adapid species developed strong sagittal crests on the upper part of the skull, which supported a strong jaw musculature. The postcranial skeleton also resembled that of the lemurs, with relatively long limbs, trunk, and tail. Despite these similarities, the adapids' anatomy was so basic that they could, in fact, be related to any group of modern primates.

The oldest known genus was *Cantius*, present in both North America and Europe. These primates were small adapids weighing between 1 and 5 kg. They retained a complete dental formula with two small vertical incisors, one prominent canine, four premolars, and three molars. The premolars and molars presented low cusps connected by crests—which suggests a mainly frugivorous diet. Like most adapids, *Cantius* was probably a diurnal primate that ran and leaped through trees using its four limbs (what is known as an arboreal quadrupedal).

Besides *Cantius*, the adapids diversified in a extraordinary way in Europe, leading to several genera of different sizes and morphologies. At the base of the adapid radiation in Europe was *Donrussellia*, which differed from *Cantius* in some details of the cheek-teeth morphology and retained, like *Cantius*, a complete dental formula with four premolars.

Joining *Cantius* and *Donrussellia* during the early Eocene was *Protoadapis*. The members of *Protoadapis* weighed between 1 and 3 kg and were in many respects similar to the earlier adapid genera. However, *Protoadapis* differed in at least one feature, having only three premolars, due to the loosening of the first one.

The omomyids were the second group of primates to coexist with the lemurlike adapids in the Eocene forests of Europe. While the adapids shared a number of features with one of the groups of extant prosimian primates, the lemurs, the omomyids resembled in many respects the second group of extant prosimians, the tarsiers. These are nocturnal primates with very large eyes and short faces that inhabit the tropical forests of Southeast Asia. Although traditionally classified with the lemuriforms as "prosimians," the tarsiers have a nose structure indicating that they are closer to the higher primates, or anthropoids (all included in the category of the haplorhines). Like extant tarsiers, the omomyids had large eyes and a relatively globular brain case. The snout was short and different from that of coeval adapids. Unlike the adapids, the omomyids are supposed to have been mainly nocturnal primates. The dentition was also different from that of the adapids, with relatively large, procumbent central incisors and small canines. Instead of the crested molars of the adapids, the cheek-teeth of the first omomyids bore sharp-pointed cusps. This indicates a less frugivorous and a more insectivorous diet than that of the adapids. The scarce postcranial skeletal remains assigned to the omomyids suggest good leaping abilities.

The oldest member of the group was *Teilhardina*, from the early Eocene in both Europe and North America. This was a primitive omomyid that shared with the primitive adapids the possession of a complete dental formula with four premolars. The omomyids were certainly much smaller than the adapids, and *Teilhardina* was among the smallest omomyids, weighing around 100 g.

THE DAWN OF THE MODERN UNGULATES

Another significant early Eocene event was the spread and diversification of the first true perissodactyls (the order that includes today's horses, rhinos, and tapirs). Perissodactyls, the odd-toed ungulates, have limbs in which the middle toe carries the main body weight.

The first true perissodactyls were small, dog-size ungulates with dental characteristics basic to both the horse and rhino groups (hippomorphs and ceratomorphs, respectively). All perissodactyls share a more specialized tarsal configuration in which the ankle bone (or astragalus) is modified to prevent any lateral movement of the foot. While the distal part of the astragalus is fixed to the navicular bone, the proximal end articulates with the tibia through a pulley-shaped surface that enables a straight rotation of the foot. This is a much better adaptation to cursoriality than the structure present in the advanced condylarths like *Phenacodus*.

The upper molars of perissodactyls also show a basic ridge-based pattern formed of a pair of transverse parallel crests that are united in the buccal side by a third external ridge (called the ectoloph). This pattern is still recognizable in present-day rhinos and tapirs and corresponds to a less omnivorous and frugivorous diet than that of the condylarths and an increasing adaptation to browsing.

The first archaic perissodactyls like *Hallensia* and *Cymbalophus* are found in Europe and probably originated from an advanced phenacodontid condylarth like *Ectocyon* (although an origin close to that of the African hyraxes has also been proposed). From the old perissodactyls like *Hallensia*, there was a sudden radiation of new forms that were among the ancestors of the horse-related group (or hippomorphs, including the horse family, or equids, and such other extinct families as the paleotherids) and the rhino-related group (or ceratomorphs, including rhinos, tapirs, and such tapirlike extinct groups as the lophiodonts).

The first true member of the horse family was *Pliolophus*. Most of the species of *Pliolophus* were once included in another genus, *Hyracotherium*, which, in fact, has proved to be a member of the paleotheres and has, therefore, been excluded from the horse's ancestry. (According to one of the reviewers of the group, Jerry Hooker, the genus *Pliolophus* includes the formerly *Hyracotherium* species "*H.*" *vulpiceps* and the American "*H.*" *pernix* but not the original *Hyracotherium leporinum* from Europe, which is, in fact, a paleothere.) The origins of *Pliolophus* can be traced from a basal perissodactyl like *Hallensia*, close to the perissodactyl stem. The dentition of *Pliolophus* and other early members of the horse family already showed a trend toward developing selenodont premolars and molars, with crescentlike ridges, as in the artiodactyl ruminants, a pattern that is superimposed on the basal perissodactyl, tapirlike morphology (this is often known as a "selenolophodont" pattern). This evolution indicates a step forward to a diet based mainly on leaves.

Although it was the first member of the horse lineage, *Pliolophus* certainly did not look like a horse. As classically stated, it had the dimensions of a medium dog ("a fox-terrier"), bearing four hooves on the front legs and three on the hind legs. The legs were short, as was the neck, and the back was still strongly curved, as in most ungulates living in dense forest.

After its dispersal to North America from Europe in the early Eocene, the equoid group split in two directions. On the one hand, the lineage leading to today's horses pursued its evolution in North America; on the other hand, during the middle and late Eocene, European equoids developed their own evolutionary trends and diversified into a variety of genera and species, all of them included in the paleotheres family.

The paleotheres were among the fossil mammals first described by the founder of vertebrate paleontology, Georges Cuvier, as early as the beginning of the nineteenth century. Cuvier regarded them as tapirlike forms, so they were frequently reconstructed with a short trunk at the end of the snout. Further studies, however, firmly established the equid affinities of the paleotheres. At the same time, a number of authors realized the peculiarities that excluded this group from the horse lineage. Early true equids like *Orohippus* already showed a trend toward the molarization of premolars and deciduous teeth, which was absent from the early European paleotheres. A reversion of the molarization pattern was even seen in some species, which developed less-molarized premolars than their ancestors. Moreover, the paleotherids had relatively shorter limbs and evolved differently in the limb proportions. Although an elongation of metapodials occurred in both paleotherids and equids, the one group showed the reverse effect of the other: while in the paleotherids the metacarpals (the intermediate bones of the forefoot) tended to be longer than the metatarsals (the intermediate bones of the hind foot), the equids showed the reverse, with shorter forefeet and longer hind feet. Elongation of the forefeet is a common feature among those ungulates adapted to browse the lower branches of trees. Therefore, the paleotheres were better suited for life in the dense forests of the early Eocene, while the equids in North America exhibited an opposite trend toward a more cursorial locomotion and a diet based on ground-level, harder vegetation.

Among the first paleotherids was the species *Hyracotherium leporinum,* once considered a true equid, which was undoubtedly close to the equid–paleotherid bifurcation. Also close to this bifurcation was *Propachynolophus,* another early paleotherid genus that in the early Eocene split into several species of different sizes, from 30 to 60 cm at the withers. From the gut contents (mostly leaves) of the closely related *Propaleotherium* of the middle Eocene at Messel, we know that these early paleotheres were mainly browsers but could also eat fruit found on the ground (grape seeds were found in the gut) (figure 2.5). Also close to the paleotherid stem was *Pachynolophus,* another archaic perissodactyl characterized by the development of higher and unbroken crests on the molars that improved the shear along the transverse crests of the upper molars at the expense of the crushing function. Forms like *Pachynolophus* advanced further in their dental adaptation to a folivorous diet.

Besides the paleotheres, another group of archaic perissodactyls, the lophiodontids, developed as an endemic group in the European archipelago and attained high diversity levels from the early to the late Eocene. Unlike the hippomorph paleotheres, the lophiodontids were a group of ceratomorphs that had a tapirlike dentition composed of brachydont molars with complete transverse crests. Because of the similarity of their teeth with those of today's tapirs, they have usually been considered related to the latter and regarded as "giant, primitive" tapirs. Consequently, most reconstructions show them with a small tapirlike proboscis. However, the features that relate them to modern tapirs are also common to other archaic perissodactyls, and the cranial anatomy does not show the characteristically retracted nasals and nasal incisions associated with the presence of a proboscis. Therefore, although tapiriform in shape, the lophiodontids were not directly related to present-day tapirs.

Actually, the lophiodontids lay close to the origin of the chalicotherids, a group of bizarre perissodactyls in which the trend to elongate the forefeet reached an extreme, developing gorilla-like morphotypes during the Miocene. In fact, the middle Eocene *Lophiaspis,* originally described as a lophiodont, is now considered a primitive, true chalicotherid (although lacking the extreme limb adaptations of its Oligocene and Miocene relatives). The most common lophiodontid was *Lophiodon,* an animal that could attain the size of a horse and that was widely found in several European localities from the early to the late Eocene. *Lophiodon* was characterized by its well-developed canines (an unusual feature in a perissodactyl), which were followed by a long diasteme. The premolars were reduced (the first one even lost) and nonmolarized. *Lophiodon* probably had a diet based on soft, nonfibrous leaves, like that of today's tapirs (plate 1). The lophiodonts made a short incursion into North America, where the genera *Paleomoropus* and *Schizotheriodes* are considered related to this group.

From the ceratomorph side, the first rhino-related forms included *Hyrachius,* a small rhino about the size of a wolf that during the Eocene inhabited a wide geographic range, from North America to Europe and Asia. *Hyrachius,* like most of the archaic rhinocerotids, lacked the quer-

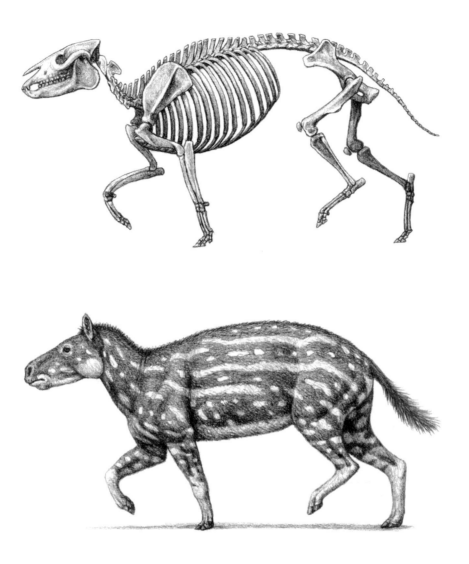

FIGURE 2.5 *Skeleton and reconstructed life appearance of the perissodactyl* Propaleotherium hassiacum

Complete, articulated skeletons of this ungulate have been found in Messel, showing primitive body proportions that are quite different from those of either the modern horses or the large, specialized paleotheres of the late Eocene. The forelimbs are shorter than the hind limbs, and the neck is remarkably short, giving the animal a rather thickset appearance. The coat pattern is based on that of juvenile modern tapirs. While adult tapirs are large animals that would gain little benefit from a cryptic coloration of this kind, a rather small, forest-dwelling animal like *Propalaeotherium* would likely have retained such a coat pattern into adulthood. Reconstructed shoulder height: 52 cm.

atinic horns that are characteristic of today's members of this group. The dentition was complete, with spatulate incisors and nonmolarized premolars. Like several archaic perissodactyls of the Eocene, *Hyrachius* bore tetradactyl forefeet and tridactyl hind feet. But unlike other ungulates of its time, it was a relatively cursorial form (figure 2.6).

In addition to the perissodactyls, the second group of advanced ungulates, the artiodactyls, or even-toed ungulates, appeared in Europe for the first time during the early Eocene. The artiodactyls are among the most successful orders of mammals, having diversified in the past 10 million years into a wide array of families, subfamilies, tribes, and genera all around the world, including pigs, peccaries, hippos, chevrotains, camels, giraffes, deer, antelopes, gazelles, goats, and cattle. They are easily distinguished from the perissodactyls because each extremity is supported on the two central toes, instead of on the middle strengthened toe. Another important particularity of these ungulates is their peculiar double-hinged ankle bone (or astragalus), which presents two symmetrical, pulley-shaped surfaces, articulating with the tibia, on the one hand, and with the tarsal navicular and cuboid bones, on the other. This evolutionary innovation has imparted a considerable advantage during fast locomotion, constraining the movement of the foot to the vertical plane and allowing leverage to be altered as required.

The oldest member of the order is *Diacodexis,* first known from the early Eocene of Europe. It was a rabbit-size ungulate with archaic, complete bunodont dentition resembling that of some arctocyonids. In contrast, its limb anatomy was advanced for its time, indicating that it was a highly cursorial animal. The lateral toes were reduced, and the metapodials (metacarpals and metatarsals) were highly elongated, with the tibia significantly longer than the femur, thus elongating the distal segment of the limbs. That the hind limbs were significantly longer than the forelimbs suggests good capability for leaping and jumping in case of danger. *Diacodexis* also had a long tail that it probably used for balance when running. The skull was lightly built and bore a long snout that enabled it to rummage through leaf litter in search of seeds, fruit,

....

FIGURE 2.6 *A lakeshore scene from Messel, with the pangolin* Eomanis waldi *and the perissodactyl* Hyrachius minimus

The environment around the paleolake at Messel corresponded to a dense, subtropical forest that would have reached right to the margin of the lake along large sections of the shore. *Eomanis waldi* would have resembled in most features a small extant pangolin, but its tail was somewhat shorter and perhaps not as useful for climbing as the long, prehensile organ of modern species. *Hyrachius,* although rather similar to *Propaleotherium* in general appearance, was somewhat larger, and details of its anatomy reveal that it was more closely related to rhinos.

worms, or even carrion. After appearing in Europe, *Diacodexis* spread over North America and Asia and diversified into a number of new species and genera.

During the early Eocene, the diversity of these archaic artiodactyls increased rapidly, and new forms like *Protodichobune* and *Cuisitherium* joined *Diacodexis*. They were similar in many respects, such as in their advanced limb configuration, quite different from that of the coeval condylarths. In contrast, their dentition exhibited a trend toward increasing selenodonty. All these archaic artiodactyl genera like *Diacodexis* and *Protodichobune* are assembled in their own family, the dichobunids, which persisted until the upper Oligocene.

THE MIDDLE EOCENE EDEN

Although the number of middle Eocene localities in Europe is quite restricted, we have excellent knowledge of the terrestrial communities of this time thanks to the extraordinary fossiliferous site of Messel, Germany. The Messel sediments were deposited at the bottom of an endorreic lake that was born as a consequence of the intense tectonic activity in northern Europe during the middle Eocene. While the progress of the rifting to the north definitively broke the terrestrial corridors between Europe and Greenland, interrupting the faunal interchange with North America, other new rifting processes led to the opening of the Rhine Valley, dividing the large central European island into two main domains to the west and east of this Rhine graben. Whereas in the western domain the marine influence increased and the sea advanced from southern England into the Paris Basin, the initial phases of development of the Rhine graben led to the deposition of large amounts of lacustrine and paludal sediments in the region today occupied by western Germany. Associated with this graben system, the Messel area formed a depression in which the continental waters from the neighboring reliefs accumulated into a system of shallow lakes and marshes. Thanks to the anoxic conditions prevailing at the bottom of one of these lakes, the Messel pit has preserved in an unusual way most of the life that surrounded its waters 49 million years ago. Leaves, branches, and fruit from the vegetational cover, several insect specimens preserved in their minute details, fish, amphibians, turtles, lizards, snakes, birds, mammals with their pelts—altogether, Messel contains a unique record of a fossil terrestrial ecosystem. Moreover, several specimens from Messel retain in their gut their last meal, providing a rare opportunity for testing the teeth-inferred dietary requirements of a number of extinct mammalian groups.

A large variety of bony fishes, such as the bowfins of the genus *Cyclurus* (close to today's *Amia*) and the large gars, about 1 m long, of the genus *Atractosteus,* populated the Messel waters. The gar was probably a fish-eater that preyed on smaller fish such as perches (*Palaeoperca, Amphiperca*), eels (*Anguilla*), and the salmonlike *Thaumaturus.* In their turn, bony fish and the other aquatic vertebrates were probably the prey of the diversified fauna of crocodiles that surrounded the Messel lake, such as *Asiatosuchus* and the alligators *Diplocynodon, Baryphracta, Allognathosuchus,* and *Pristichampsus.* Large boids (the 2-m-long *Palaeophyton*) and small pipe snakes of about 50 cm frequented these waters, as they do today in the rivers of tropical rainforests.

A dense canopy forest surrounded Messel lake, formed of several tropical and paratropical taxa that today live in Southeast Asia (figure 2.6). Water lilies (Nymphaeaceae), reeds (Typhaceae, Sparganiaceae), sedges (*Carex,* Restionaceae), and swamp cypresses (*Taxodium*) occupied the swampy areas close to the margins. A variety of lianas of the moonseed and grape families (Menispermaceae and Vitaceae, respectively) populated the dense canopy in the interior of the forest. The humid, stratified evergreen forest was formed of several large-leaved trees that today are found in the tropical and paratropical domains, like palms, laurel trees (*Laurus*), cinnamon and camphor trees (*Cinnamomum*), avocado pears, and a number of paratropical conifers that survive in some Chinese forests (*Cathaysia, Sciadopytis, Glyoptostrobus, Cephalotaxus*). Small climbing bushes (Rutaceae and bushes of the tea family like *Polyspora*) made up the undergrowth of these forests, which was probably covered by leaves, small twigs, and fruit (mainly grapes and nuts). On the ground, a variety of insects, such as burrowing bugs and ants, contributed to the reworking of this vegetational cover. A number of these insects reached large size—for example, some giant ants (queens with a wingspan of 16 cm) and cockroaches (up to 5 cm long). Beetles of many kinds occupied the open spaces and the grounds of these forests, including chick-beetles, weevils, jewel beetles, dung beetles, leaf beetles, as well as a variety of diurnal and nocturnal ground beetles.

THE MAMMALIAN FAUNA OF MESSEL

Given the extraordinary variety of beetles and other insects that inhabited the Messel forest, it is not surprising that most of the mammalian elements from this site were small, insectivorous taxa like the hedgehog-related *Macrocranion,* which we described in chapter 1 as an example of primitive erinaceomorph insectivores. *Macrocranion* was a small, nocturnal ground form with small eyes, large ears, and a mobile snout; it lacked the characteristic bristly fur of today's hedgehogs. More robust and hedgehoglike was *Pholidocarpus,* a member of the extinct family of the amphilemurids, which bore a dense fur on its back made of acute, relatively long bristles. But this bristly fur excepted, the general shape of *Pholidocarpus* had little in common with that of modern hedgehogs. The forehead was covered by a horny plate, or leathery callus, unique among the erinaceomorph insectivores. The tail was also covered by a tube formed by small bony scales, as in today's spiny mouse of Africa (*Acomys*). This association of characteristics is difficult to interpret but suggests that *Pholidocarpus* was a ground-dweller that used its horny forehead plate for burrowing in the ground. The amphilemurids attained a certain diversity during this period, and forms like *Alsaticopithecus* and *Amphilemur* were common at other Eocene sites.

Several families of insectivore-like placentals were represented at Messel as well. We have already referred to the leptictid *Leptictidium,* which in the middle Eocene included at least three species of different sizes (figure 2.7). All three bore the strange features of the genus: a long, slender snout that ended in a short trunk, extremely shortened forelegs, elongated hind legs, and a long tail, with more than forty vertebrae, that was used for balancing during fast running and jumping. The gut contents of the Messel specimens indicate an insectivorous/omnivorous diet based on large insects and small vertebrates.

FIGURE 2.7 *Reconstructed life appearance of the leptictid* Leptictidium nasutum

Based on complete skeletons from Messel, we know that this animal had a general appearance vaguely resembling that of a large modern elephant shrew, but the disproportion between fore- and hind limbs was even more marked, indicating habitual bipedal locomotion. The hind limbs and vertebral column, however, do not show marked saltatorial adaptations, unlike those of many modern rodents and marsupials with long hind limbs, so it seems that at least during slow gaits, the animal would walk bipedally rather than jump. This is a unique construction among mammals, and in functional terms is a solution similar to that found among small bipedal dinosaurs. Reconstructed rump height: 21 cm.

The apatemyids at Messel included *Heterohyus,* a genus that bore the family's cranial and dental adaptations, with strong, procumbent incisors. Moreover, the Messel specimens of *Heterohyus* give us an idea of the postcranial anatomy of this specialized group of archaic placentals. The first surprising feature is the small body relative to the skull, as well as the long tail; both features suggest an arboreal lifestyle. Even more striking is its hand anatomy, in which the second and third fingers were much longer than the others. A similar adaptation is found in today's lemur *Daubentonia,* which uses its long fingers to winkle out insect larvae that are located in wood. *Heterohyus* probably used its long fingers for the same purpose, after tearing the bark open with its large incisors.

Besides leptictids and apatemyids, otterlike pantolestids were present in Messel as well. The full anatomy of the specimens of *Buxolestes,* briefly described in chapter 1, confirms the aquatic habits of these archaic placentals, with their strong and highly mobile forearms and powerful tail formed of vertebrae with large transverse processes. Other features, however, indicate that they were also good land runners. The presence of large bony claws suggests that, like today's otters, they were able to build underground dens. This mixed, semiaquatic–semiterrestrial lifestyle is also reflected in the gut contents of the two specimens found: while the gut of one specimen contained scales and fish bones, the second contained only fruits and seeds.

In addition to the insectivores and insectivore-like placentals, the most common insect-eaters in Messel were bats, represented by several well-preserved specimens. Among them were the large, archaic *Archaeonycteris,* as well as the small, highly specialized *Palaeochiropteryx,* which with its broad wings practiced a reduced wing-loading flight close to the ground or the water surface (figures 2.3 and 2.4). In contrast was the highly specialized *Hassianycteris,* the largest Messel bat, with a wingspan of between 40 and 50 cm. Unlike those of the small *Palaeochiropteryx, Hassianycteris*'s wings were relatively narrow, indicating high wing-loading. Most probably, *Hassianycteris* flew fast in the open spaces of the high arboreal strata of the Messel forest, as the mastiff bats of the family Molossidae do today.

Rodents and primates at Messel basically included the same groups that were present during the early Eocene. Among the rodents, the most common forms were the large ailuravids of the genus *Ailuravus* and the small ischyromyids of the genus *Microparamys.* But there was also a third autochthonous genus: *Massillamys. Massillamys* was intermediate in size between *Ailuravus* and *Microparamys* and, like the latter, had short legs. Its limb proportions were closer to those of some digging rodents, like modern voles. Among the primates, the lineage of the lemurlike *Microadapis* continued in Messel with *Europolemur,* an adapid genus that differed from the former only in minor dental details (in fact, some authors consider them the same genus). The tarsiform-like omomyids are absent from the Messel record, but we know from other localities of similar age that they were part of middle Eocene terrestrial ecosystems.

A small species of the hyaenodont *Proviverra*, of about 20 cm, represented the creodonts at Messel (plate 1). In the middle Eocene, these small forms were still far away from the hypercarnivorous specializations of the late Eocene hyaenodontids, maintaining a mixed diet based on small vertebrates and insects and other invertebrates. However, two genera, *Parodectes* and *Miacis*, represented the true carnivores of the family Miacidae. *Parodectes* was about 60 cm long, a plantigrade form with a long tail that it probably used for balance when running (figure 2.1). The species of *Miacis* from Messel was much smaller, about 20 cm. Both had strongly developed clavicles. The clavicles are vestigial in most modern carnivores, and their retention in these archaic miacids indicates that they were agile climbers that probably preyed on smaller arboreal mammals.

With the possible exception of the paroxyclaenid *Kopiodon*, the condylarths are entirely absent from the Messel record (however, we must remember that some authors consider this genus related to the pantolestids and not to the arctocyonids). *Kopiodon* was an arboreal mammal, about 1 m long, that was especially well suited to climbing, as revealed by the retention of the clavicles and the presence of large bony claws at the ends of the fingers. As in other paroxyclaenids, its complete dentition indicates that *Kopiodon* was mainly a fruit-eater.

The truly advanced condylarths like *Phenacodus* were still present in Europe during the middle Eocene (plate 1). By then, however, the perissodactyls were the most common ungulates of the terrestrial ecosystems, represented by the equoid paleotherids, the tapirlike lophiodonts, and the archaic rhino *Hyrachius*. These three early perissodactyl groups are present at Messel, as well as the archaic, stem hippomorph *Hallensia*. Two species of the genus *Propaleotherium* represented the paleotheres. This genus was in fact very close to *Propachynolophus* and can be considered its direct descendant (if both are not, in fact, the same thing). In the exceptionally well preserved specimens of Messel, metacarpals and metatarsals were almost the same length, thus indicating that this genus was still very close to the equid–paleotherid split (some authors, in fact, consider both *Propachynolophus* and *Propaleotherium* as true equids, placed before that divergence). The smaller species from Messel, *P. parvulum*, was the size of a fox terrier, around 30 to 35 cm at the withers. A second species, *P. hassiacum*, was somewhat larger, about 60 cm at the withers (following the dog metaphor, the size of a German shepherd) (figure 2.5). The representation of the other perissodactyl groups in Messel was limited, the tapirlike lophiodonts being represented by some scarce remains that cannot be recognized at the generic level. From the rhino group, *Hyrachius* persisted from the early Eocene.

Among the artiodactyls, increasingly selenodont forms like *Messelbunodon* and *Aumelasia* represented the dichobunids. *Messelbunodon* was the size of a dog and, like *Diacodexis,* was a fast runner with an advanced limb configuration, strong hind limbs, reduced lateral toes, and a long tail. *Aumelasia* was close to *Messelbunodon* and differed only in some details of its dentition.

Besides the dichobunids, the middle Eocene forest of Messel records the appearance of a second type of artiodactyl: the haplobunodontid *Masillabune.* The haplobunodontids were related to a group of archaic artiodactyls (Bunodontia) that also included the choeropotamids and cebochoerids. In the middle and late Eocene, they began to develop bunodont molars and shorter limbs, as in today's suiforms (the group that includes pigs and peccaries). This poses an interesting problem, since the basic anatomy of suiforms, with femur and tibia of about the same length and relatively short legs, is apparently more primitive than that of the earliest dichobunids and fits better as the basal locomotive design for all the artiodactyls. If the advanced limb configuration of *Diacodexis* were the real primitive condition of the order, this would mean that the generalist condition of the suiforms is a secondarily derived feature and a reversion in the trend toward increasing cursoriality.

In any case, by the middle Eocene, a basic bifurcation was already established within the artiodactyls, with one group bearing selenodont dentition and slender, elongated limbs adapted to fast running (Selenodontia), and a second group of generalist forms, bearing bunodont dentition and more robust, relatively shorter legs (Bunodontia, including some old Eocene families like the choeropotamids and the pig-related suiforms).

THE MESSEL ENIGMA

Thanks to the extraordinary preservation of the Messel material, this locality has provided information not only about the common elements present in other localities of similar age, but also about the rare forms that are usually absent. Knowledge of these strange elements has considerably enlarged our views of the middle Eocene terrestrial faunas of Europe.

The pangolins of the genus *Eomanis* were among the most peculiar members of the Messel fauna. Today's pangolins are medium-size, ant- and termite-eating mammals that live in the forests and savannas of Africa and southern Asia. They have large claws and a long, narrow tongue, like the South American anteaters. However, their most distinctive feature is their unique covering of horny scales over the head, body, outer surfaces of the limbs, and tail. Similarly, *Eomanis* was a plump animal about 50 cm long that, despite the difference in some 49 million years, showed characteristics that are recognizable in its modern relatives (figure 2.6). The only differences are the absence of scales in the tail and a lesser capability to roll itself up, as the *Manis* of today do as a defense mechanism.

But even more surprising than the first appearance of pangolins in Messel was the unique finding of the edentate *Eurotamandua*. The edentates include today's anteaters, sloths, and armadillos of South America. All of them are, in fact, primitive therians that retain a common urinary and genital duct, internal testes in the males, and a primitive divided womb in the females. Despite these primitive characteristics, they have particular specializations, such as the presence of additional articulations between the lumbar vertebrae (called xenarthrales) that provide a reinforcement for digging. Today's edentates also differ from other mammals in having rather simple skulls with reduced dentition, which is completely lost ("edentate") in the anteaters.

The middle Eocene *Eurotamandua* from Messel was similar to today's *Tamandua* from South America: the collared anteater (figure 2.8). Unlike its close relative the giant anteater, which is exclusively a ground-dweller, the collared anteater can also live in trees. *Eurotamandua* had similar dimensions (about 1 m long) and probably had a similar lifestyle. The presence of *Eurotamandua* in Messel poses an interesting zoogeographic problem, since the edentates are basically known from South America, and only one other genus, *Ernanodon*, has been found outside this continent, in the Chinese Paleocene (but the latter genus was more closely related to the sloth/armadillo group than to anteaters).

This mixture of European and South American elements in Messel is not restricted to placental mammals. Among the birds, the giant diatrymas, which we mentioned in chapter 1, were still present during the middle Eocene at Messel and Geiseltal, represented by the species *Diatryma geiselensis*. Besides the diatrymas, a second large predator bird, the phorusracid *Aenigmavis*, was present. Like the diatrymas, the phorusracids were giant, flightless birds with powerful beaks and strong hind limbs armed with long claws. Probably related to the gruiforms, these avian predators attained gigantic dimensions during the Oligocene and Miocene in South America, where they probably played a role similar to that of the diatrymas in Europe. The size of a large rooster, *Aenigmavis* was smaller than the South American phorusracids but certainly related to them.

A third South American–Messel connection existed among the crocodile community. Besides the crocodile *Diplocynodon* and the alligators *Asiatosuchus*, *Allognathosuchus*, *Pristichampsus*, and *Baryphracta*, the Messel beds have yielded the remains of a terrestrial meat-eating crocodile, *Bergisuchus*, which is related to the archaic mesosuchian crocodiles. The

FIGURE 2.8 *Reconstructed life appearance of the anteater* Eurotamandua joresi

Although the relationship between *Eurotamandua* and the extant anteaters of South America is still a matter of debate, the structural similarities are very close and reflect a similar adaptation for tearing into the nests of ants and termites using well-muscled forelimbs armed with large claws. Reconstructed total length: 90 cm.

roots of this group can be traced to the Mesozoic of the Gondwanan continents (South America and Africa), so its presence in Messel is further evidence for a zoogeographic connection to South America.

In fact, it seems highly probable that the South American elements like *Eurotamandua*, the phorusracids, and the mesosuchian *Bergisuchus* reached Europe through the African continent. Although this idea might seem somewhat strange, during the early and middle Eocene the South Atlantic had not yet reached its present dimensions, and communication between South America and Africa was still possible. The presence of elements of possible African origin in Messel supports this faunal connection. For instance, we have mentioned the didelphoids *Peradectes* and *Amphiperatherium,* both present in Messel. The original distribution of these opossums was restricted to the southern continents and, more specifically, in the case of the didelphids, to South America. The finding of a didelphid in the early Eocene of northern Africa suggests that these marsupials could have entered Europe directly from South America, across an African–Iberian corridor. Also, the first prosimian primates are known from the Paleocene of northern Africa, suggesting that they could have dispersed into Europe by following this Iberian corridor. This could also have occurred with the first modern ungulates, perissodactyls and artiodactyls, although in this case there is no direct evidence. With regard to the perissodactyls, the origin of the paleothere populations in Europe is still a subject of debate. However, a direct relationship between the archaic perissodactyls and the African hyraxes has been proposed, on the basis of a number of shared characteristics. Moreover, the oldest *Hyracotherium* is first known from southern Europe (Silveirinha) before its dispersal into northern Europe. Similarly, the first artiodactyls of the genus *Diacodexis* exhibit a south–north cline through time. The connection between Africa and southern Europe could, therefore, explain the sudden appearance in the early and middle Eocene of Europe of such modern groups as artiodactyls and perissodactyls, as well as the presence of a number of South American elements that could have used the African–Iberian corridor.

THE BEGINNING OF THE END

At the end of the middle Eocene, things began to change in the European archipelago. Several late Paleocene and early Eocene survivors had become extinct, including the last multituberculates, the plesiadapiforms, and the arctocyonid and hyopsodontid condylarths, as well as the tillodont *Esthonyx* and other strange Paleocene remnants. Some advanced phenacodontids (*Phenacodus, Almogaver*) survived for some time in southern Europe, such as in the Tremp Basin in Spain, before being fully replaced by the perissodactyls and artiodactyls during the late Eocene (plate 1). The last part of the middle Eocene saw a clear change in the structure of the herbivore community as specialized browsing herbivores like the lophiodonts replaced the small to medium-size omnivorous/frugivorous archaic ungulates of the early Eocene and became the dominant species. At this time, this group of large tapirlike perissodactyls diversified into a variety of genera and species, including *Lophiodon* (which had arisen during the early Eocene), *Atalonodon,* and *Paralophiodon.* The lineage of archaic rhino-related hyrachids continued in Europe as well, represented by *Chasmotherium,* a species formerly included among the lophiodontids.

This change affected not only the large herbivores, but also the smaller ones. While some middle Eocene holdovers, like the ailuravids and the large ischyromyids of the genus *Plesiarctomys,* continued without great variations, most of the small species of the genus *Microparamys* declined and became extinct. However, some *Microparamys* species persisted, and their descendants included the glirids and theridomyids, two of the most important families in the later Eocene and Oligocene. In both cases, there was a trend toward lophodont molar designs composed of a succession of transverse crests. In the glirids, these crests tended to be low and numerous. This group of rodents includes today's dormice, which are a common element of the small mammalian faunas of the Palearctic. The name dormice refers to their capability to hibernate during the cold season in temperate regions. They are mainly arboreal animals with a diet based on fruits and seeds. Their molars are usually brachydont and formed of a succession of low, parallel ridges. The first glirids like *Eogliravus* and *Gliravus* still had a simple dental pattern, composed of few crests, as in today's *Eliomys;* this pattern indicates a wide diet based not exclusively on fruits.

The second group of rodents, the theridomorphs, initially exhibited a basic dental pattern not very different from that of some advanced species of *Microparamys,* like *M. chandoni.* However, the theridomorphs went further in the development of new ridges uniting the main cusps of the teeth. From the middle Eocene *Protadelomys,* the theridomorphs soon split into two main groups with opposite evolutionary trends. On the one hand, the pseudosciurids developed a squirrel-like dental pattern, with low-crowned, brachydont teeth and a bunodont pattern, adapted to crushing fruits and soft vegetables. On the other hand, the theridomyids exhibited progressively hypsodont molars, which displayed a characteristic crown design based on five main transverse ridges, as in modern guinea pigs, chinchillas, and other South American endemic rodents living in open environments. This pentalophodont pattern (as it is known) indicates a trend toward a more fibrous diet based on harder vegetables. We know from the postcranial anatomy of the well-known *Pseudoltinomys* that some of these theridomyids were open-environment ground-dwellers, with elongated hind limbs adapted to running and jumping on the hard ground (figures 3.6 and 3.8).

Among the insectivore-like placentals, several archaic groups such as the leptictids declined and finally became extinct. In contrast, new families appeared and expanded among the true insectivores. For example, the nyctitherids, a group that originated in North America in the middle Paleocene, attained in the Eocene a Holarctic distribution. According to the tooth morphology, characterized by a W-shaped pattern, the origin of shrews (soricids) and moles (talpids) can be traced back to some primitive members of the nyctitherids. Therefore, by the middle Eocene the main groups of insectivores, the erinaceomorphs (hedgehoglike) and the soricomorphs (shrewlike), were already present, represented by the adapisoricines and the nyctitherids.

Among the predator guild, the hyaenodonts continued as the dominant meat-eaters. However, some scattered, although highly significant, entries are recorded at this time, such as *Simamphicyon helveticus*, an early member of the amphicyonid family. The amphicyonids were archaic carnivores characterized by short limbs, a robust appearance, a long tail, and a generalized dentition that retained the third molars. Their robust anatomy, associated with a doglike dentition, led some authors to coin the popular name bear-dogs, although the actual affinities of the group are still matters of discussion. Because of their dentition, they have been more frequently regarded as canids (the wolf and dog family) than as ursids (the bear family), but some particular traits, such as the carotid anatomy, would indicate a closer relationship to the latter. Anyway, there is today a tendency to place them in their own family, the amphicyonids. Scarcely represented in the late Eocene of Europe, they became one of the most important carnivore groups during the Oligocene and Miocene.

These changes among the mammalian faunas were most probably a response to the major tectonic transformations occurring at that time and the associated environmental changes. During the middle Eocene, the Indian plate collided with Asia, closing the Tethys Sea north of India. The collision of India and the compression between Africa and Europe formed an active alpine mountain belt along the southern border of Eurasia. In the western Mediterranean, strong compression occurred during the late Eocene, with crustal shortenings and emplacement of nappes to the north of the Iberian Peninsula, leading to the final emergence of the Pyrenees. To the south of the Pyrenees, the sea branch between the Iberian plate and Europe retreated, and fluvial and paludal sediments began to fill the trough separating both domains. A number of significant late Eocene fossiliferous localities like Sossís and Roc de Santa were formed in the Tremp Basin, south of the Pyrenees. In central Europe, the rifting process continued during the late Eocene, associated with intense volcanism. The Rhine graben opened to the south, being filled by brackish sediments and potash salts. To the east, the Turgai Strait, as a shallow seaway, still connected the warm waters of the Tethys with the Polar Sea. However, because of the northward motion of the African plate, the Indo-Pacific marine circulation became progressively restricted, increasing the contrast between sea-surface temperatures and affecting in this way the late Eocene climates.

THE LATE EOCENE FAUNAS

The changes initiated in the middle Eocene continued into the late Eocene. Among rodents, the theridomorphs exploded into a variety of genera and species. At the same time, a high degree of regional differentiation appeared among these late Eocene theridomorph faunas. Thus the brachydont pseudosciurids, well adapted to the ingestion of fruits and soft vegetables, dominated the area east of the Rhine graben with a number of genera such as *Suevosciurus, Treposciurus,* and *Sciuroides.* West and south of the Rhine graben, the lophodont theridomyids, like *Theridomys, Patriotheridomys, Paradelomys,* and *Blainvillimys,* dominated the rodent communities. Some elements of this group, like *Pseudoltinomys* and *Elfomys,* began a precocious evolution toward developing high-crowned, hypsodont molars, indicating their adaptation to a diet based on hard and fibrous vegetables. The arboreal leaf-eater ailuravids like *Ailuravus* persisted, while only the large *Plesiarctomys* represented the ischyromyids.

In addition to the theridomorph rodents, the small-mammal communities were basically composed of nyctitherid insectivores (*Saturninia*), apatemyids (the medium-size *Heterohyus*), and numerous primates. Among the prosimian primates, there was a quick diversification of the tarsiform omomyids into several genera and species, all of them included in the autochthonous family Microchoeridae. The microchoerids appeared for the first time in the middle Eocene with *Nannopithex,* a small omomyid of about 125 to 170 g. This animal had a large, procumbent first incisor that resembled the large canine of today's tarsiers. The other incisors were absent or much reduced and were followed by pointed premolars and molars, which indicates an insectivorous diet. The skull was short, with a narrow snout and large orbits suggesting a nocturnal life. During the late Eocene, the microchoerids diversified into a number of genera, such as *Pseudoloris, Microchoerus,* and *Necrolemur.* The smallest form, *Pseudoloris,* was in many respects comparable to *Nannopithex,* although even smaller (between 50 and 120 g). The other late Eocene microchoerids were, however, larger than *Pseudoloris, Necrolemur* weighing around 300 g and *Microchoerus* being close to 1 kg (the species *M. erinaceous* reached 1.7 kg) (figure 2.9). The shape of their molars, with low, rounded cusps and crenulated enamel, also suggests a diet based on fruits rather than insects.

Besides the nocturnal tarsiform microchoerids, the diurnal lemuriform adapids were common in late Eocene forests, represented by *Adapis* and the larger *Leptadapis. Adapis* is among the best-known Eocene

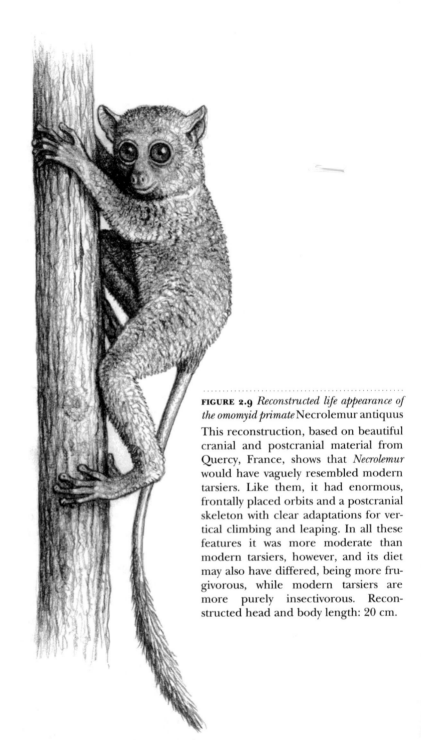

FIGURE 2.9 *Reconstructed life appearance of the omomyid primate* Necrolemur antiquus

This reconstruction, based on beautiful cranial and postcranial material from Quercy, France, shows that *Necrolemur* would have vaguely resembled modern tarsiers. Like them, it had enormous, frontally placed orbits and a postcranial skeleton with clear adaptations for vertical climbing and leaping. In all these features it was more moderate than modern tarsiers, however, and its diet may also have differed, being more frugivorous, while modern tarsiers are more purely insectivorous. Reconstructed head and body length: 20 cm.

primates, thanks to the several skulls and postcranial elements of *A. parisiensis* recovered from the fissure infillings of Quercy in France (figure 2.10). It was a medium-size adapid of about 2 kg that showed the typical features of the family, with a low, broad skull, small eyes, and prominent sagittal crests. The upper jaw was armed with spatulate central incisors, while premolars and molars were relatively narrow and

FIGURE 2.10 *Reconstructed life appearance of the adapid primate* Adapis parisiensis
While many genera of Eocene primates would have moved among the branches with powerful leaps, European adapines like *Adapis* and *Leptadapis* were somewhat slower animals, able to climb efficiently but less well equipped for leaping than *Necrolemur* or the notharctines (American members of the Adapidae). Postcranial remains of *A. parisiensis* found at Quercy show that this animal had relatively shorter hind limbs than typical leaping prosimians, as well as a shallower distal femur and a short calcaneum, all adaptations to a wide range of rotation at the hind limb for climbing, rather than to a springlike propulsion for jumping.

bore well-developed shearing crests. These characteristics indicate that unlike the small microchoerids, such as *Pseudoloris*, *Adapis* was folivorous. In its turn, *Leptadapis* was a large adapid that could attain more than 8 kg in the species *L. magnus* (figure 2.11). The males of *Leptadapis* had larger skulls and longer canines than the females, this being an almost unique case of sexual dimorphism in prosimians.

Among the creodonts, the late Eocene coincides with a sudden diversification of the group, with several new forms like *Hyaenodon, Pterodon, Cynohyaenodon, Paracynohyaenodon, Quercytherium,* and *Paroxyaena* entering Europe at this time. Most of them, like the hyaenodonts *Hyaenodon* and *Pterodon*, were probably of Asian origin (figure 2.12). The hyaenodont *Hyaenodon* was a hypercarnivorous form that possessed highly sectorial teeth with well-developed slicing blades like those of the big cats. It bore long canines, which were slightly larger in the males, thus suggesting limited sexual dimorphism. Unlike the oxyaenids and archaic hyaenodonts, *Hyaenodon* was a cursorial form, with elongated, slender limbs that ended in digitigrade feet. However, its compact body and limited pronation-supination capabilities indicate that this animal was not a specialized pursuit carnivore like extant canids, but a generalized predator (plate 2). Quite different was the carrion-eater *Quercytherium*, which developed large, crushing molars, like those of today's hyenas, as an adaptation to bone-cracking.

In contrast to the high diversification among hyaenodonts, late Eocene carnivores included only a few genera. The small, arboreal miacids still included *Paramiacis.* A second genus, *Cynodictis,* joined these miacids in the late Eocene. *Cynodictis* was in many respects close to the miacid stock but already displayed some advanced characteristics that indicate its relation to the canids. Despite their familiarity to us, the members of the dog family are generalized carnivores close to their ancestral stock. The transition from the early miacids to the first canids took place through forms like *Cynodictis* by the development of teeth with less puncturing cusps and the loss of the last molars. After *Cynodictis,* the evolution of the true canids occurred in North America, while in Europe the bears (ursids) and bear-dogs (amphicyonids) became the dominant mesocarnivores.

The paleotheres became the dominant perissodactyls in the late Eocene, replacing the tapirlike lophiodontids. The old, stem paleothere genera, like the brachydont *Pachynolophus,* also declined, and more advanced forms, like *Plagiolophus* and *Paleotherium,* spread over Europe. The first representatives of this new wave of paleotherids were relatively small, not differing in size from their middle Eocene relatives but already showing increasing browsing adaptations, such as more-elongated neck vertebrae and longer forefeet (figure 2.13). During the late Eocene, the paleotherids radiated into a wide array of species of different

FIGURE 2.11 *Skull and reconstructed head of the adapid primate* Leptadapis magnus

Well-preserved skulls from the site of Quercy show that the masticatory apparatus of *Leptadapis* was well adapted to a folivorous diet: the teeth were crested for cutting through leaves, the temporal crest for insertion of the temporalis muscles was enormously developed, and the zygomatic arch was high and strong to withstand the forces exerted by the chewing musculature. Total length of skull: 12 cm.

FIGURE 2.12 *Skull, musculature, and reconstructed head of the hyaenodontid* Pterodon dasyuroides

Like so many other mammals of the European Eocene, *Pterodon* is best known from fossils found at Quercy. The skull and mandible of this creodont were vaguely dog-like, with a long muzzle armed with teeth that allowed it both to cut meat and skin and to crush bone with moderate efficiency. The arrangement of the muscles of mastication, as inferred from the insertion areas in the skull and mandible, indicates an advanced adaptation of the masticatory apparatus for the sectorial (that is, cutting) function of the cheek-teeth. Total length of skull: 24 cm.

sizes and cursorial adaptations, complicating in a extraordinary way the classification of this group. The succeeding species that appeared at this time, like *Paleotherium medium,* were larger and showed a trend toward the molarization of the premolars and the development of higher-crowned molars with a fully selenolophodont, crescentlike design. At the same time, the feet became longer and slenderer, adapted to faster locomotion. This is an indication of the extension of more open, dry conditions in Europe during the late Eocene. Some paleotheres achieved at this time considerable dimensions; for example, *Paleotherium magnum* was a horse-size species with high-crowned molars and a rela- tively long neck (figures 2.14 and 2.15; plate 2). In the late Eocene, the Iberian Peninsula maintained a certain degree of isolation, with the open Atlantic waters still covering the western part of the Ebro Basin. Consequently, an autochthonous paleothere fauna developed, com- posed of various genera like *Paranchilophus, Franzenium,* and *Cantabroth- erium,* which mimicked the trends observed to the north and east of the Pyrenees. For instance, *Cantabrotherium* was a large paleothere, compa- rable in size to *P. magnum* but having developed, among other features, much higher-crowned, selenolophodont molars.

THE LATE EOCENE ARTIODACTYL RADIATION

Despite the high specific diversity of the paleotheres, the most successful and innovative ungulate group at the beginning of the late Eocene was the artiodactyls, which from a basal dichobunid stock radiated into a variety of new families and morphotypes. The dichobunids not only survived but even attained a certain diversification, with new genera like *Dichobune* and *Mouillacitherium.* The piglike morphotype, with bunodont molars and short legs, also diversified into a number of genera and families, such as the still persisting haplobunodontids (*Haplobunodon*), the cebochoerids (*Cebochoerus*), and the most common choeropotamids (represented in several localities by *Choeropotamus*). Among the buno- dont artiodactyls, however, the most significant event was the entry of the anthracotheres, a group of Asian origin. The first anthracotheres, like *Raghatherium, Lophiobunodon,* and *Anthracobune,* were bunodont, piglike forms larger than a rabbit and smaller than a wild boar. During the late Eocene, they attained larger sizes and developed selenodont molars; this dentition would indicate a deviation from the original om- nivorous diet to a browsing regime. However, unlike today's suiforms, the anthracotheres retained five digits in the forefeet and four in the hind feet (although the lateral ones were more reduced). This was a unique combination of ruminant-like, selenodont dentition coexisting with a generalist, pig- or hippolike shape, with relatively short, stout legs and still functional lateral digits (in fact, a combination close to that of

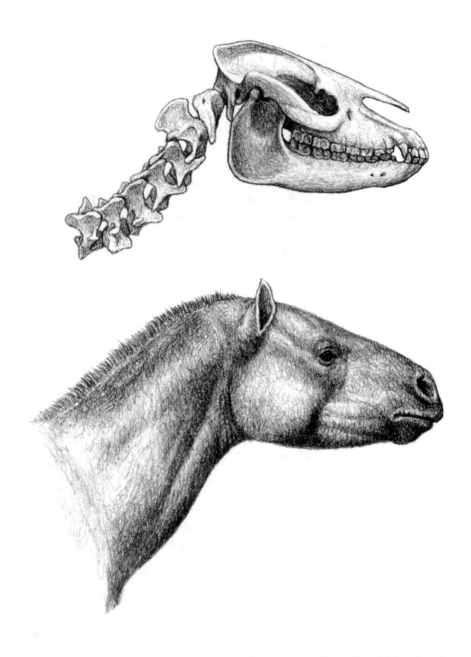

FIGURE 2.13 *Skull, cervical vertebrae, and reconstructed head and neck of* Paleotherium muehlbergi

This reconstruction, based on remarkably complete cranial and cervical material from the site of Saint-Etienne-de-l'Olm, France, shows the traits of a large, derived paleothere. Although the skull appears vaguely horselike, the orbits are set much farther forward than in true horses, due in part to the greater development of the temporalis muscles, which require longer temporal fossae. The neck is elongated, although not so much as in the larger *Paleotherium magnum.*

today's hippos). This family succeeded during the late Eocene and the Oligocene in Europe and persisted in Africa up to the Pliocene, when it probably gave rise to the hippopotamuses. During the late Eocene and the early Oligocene in Europe, the most common anthracothere was *Elomeryx,* a medium-size form that, like other anthracotheres, probably had a semiamphibious lifestyle, comparable to that of African hippos (plate 3).

The late Eocene also records a wide array of advanced artiodactyls that developed more selenodont dentition as well as cursorial adaptations. Their origin lies close to an advanced dichobunid like *Mouillacitherium,* and they soon diversified into a number of families, such as the dacrytherids, cainotherids, anoplotherids, and xiphodontids. A common feature was a trend toward developing a fully selenodont dentition, somewhat approaching that seen in the tylopods (camels and llamas) and ruminants. In association with these changes in dentition was a trend to elongate the anterior part of the skull, resulting in the development of a diasteme between canines and premolars. At the same time, the incisors became procumbent, as an adaptation to browsing, while the lower canines acquired an incisor shape.

The most primitive family of this assemblage, still close to its dichobunid origins, was the dacrytherids, represented by *Dacritheryum* and *Tapirulus.* These genera retained a complete, moderately selenodont dentition, in which the cusps were still recognizable and a diasteme was not yet formed.

Close to the dacrytherids were the anoplotherids, which again retained a complete dentition without diasteme but bore more selenodont teeth. Although still short-legged, this family exhibited a trend toward increasing cursoriality, as shown by its almost bidactyl limbs in which the lateral digits were absent or much reduced. Therefore, *Diplobune* had three digits on each foot, while *Anoplotherium* retained only two.

Also related to the anoplotherids were the cainotherids, a group of small, selenodont artiodactyls that appeared in Europe during the late Eocene and persisted in this continent up to the end of the early Miocene, 14 million years ago. The first cainotherids can be traced back to the beginning of the late Eocene, rooted in the small *Robiacina.* Later, they diversified into a number of genera that persisted into the Oligocene, such as *Oxacron* and *Paroxacron.* They were rabbit-size artiodactyls with a complete, highly selenodont dentition. The skull was small, with a short snout and closed orbits that were placed in a middle position. A peculiar characteristic of this group was the timpanic bullae, which were very large, as in today's small mammals adapted to open, subdesert environments. This advanced cranial morphology contrasted with the animal's postcranial anatomy, which was rather archaic. The cainotherids retained four nonreduced digits on each limb (although the

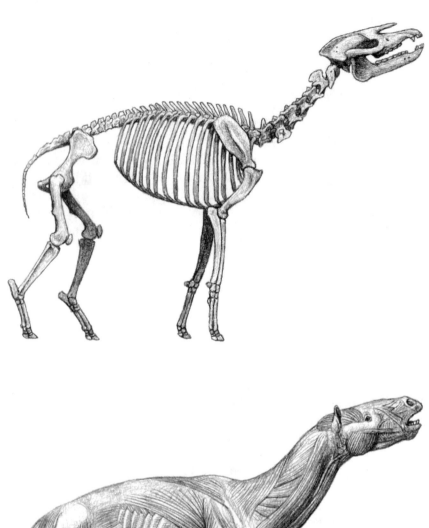

lateral ones were shorter), which ended in long claws, as in modern rabbits. The hind limbs were longer than the forefeet, indicating good capabilities for running and jumping. Because of their small size and similar locomotor adaptations, the cainotherids have often been regarded as an ecological equivalent of today's rabbits and hares. However, the foot imprints left by the early Miocene *Cainotherium* found at Salinas de Añana in Alava, Spain, suggest a kind of locomotion close to that of the small ruminants and quite different from that of the lagomorphs.

In contrast to the cainotheres, the xiphodontids exhibited much different locomotive adaptations. The xiphodontids were highly selenodont, with a complete dentition in which the premolars were elongated and developed shearing surfaces. The gazelle-size *Xiphodon* was a highly cursorial form that had elongated, slender bidactyl limbs highly adapted for fast running (plate 2). As in the case of other artiodactyls of this time, this adaptation suggests a trend toward the development of open spaces and the reduction of the closed forests of the late Eocene. Because of some features of their dentition and postcranial anatomy, the xiphodontids have been regarded as early tylopods, which at that time began to develop in North America. However, the xiphodonts look like archaic camels not through derived features but through primitive ones also shared with other groups.

Despite the advanced limb anatomy of the xiphodontids, the most derived group among the late Eocene assemblage of archaic, preruminant artiodactyls was the amphimerycids. This family, composed of a number of genera like *Pseudamphimeryx* and *Amphimeryx,* demonstrated to an extreme the trend to develop selenodont cheek-teeth and to reduce the lateral digits. The amphimerycids not only had lost the lateral digits (as in the closely related xiphodontids) but already had fused the two remaining metatarsals into the so-called cannon bone, which is present in most ruminants.

....

FIGURE 2.14 *Skeleton and reconstructed musculature of* Paleotherium magnum

Although traditional reconstructions of *Paleotherium magnum* depicted it as resembling a modern tapir, down to the detail of having a short trunk, the almost complete skeleton from Mormoiron, France, clearly shows that it had a rather unique appearance. It had relatively long legs, and the forelimbs were noticeably longer than the hind limbs, while the reverse is true for tapirs. Also unlike tapirs, *P. magnum* had an elongated neck, which, coupled with its sloping back, would give its body a vaguely okapi-like appearance, although more robust. While there was some retraction of the nasals, we see none of the extreme specializations of the nasal area found in tapirs and other animals with a proboscis, so it is clear that paleotheres did not have a trunk. All in all, the body proportions of *P. magnum* corresponded to those of a browser that fed on relatively high branches, being able to reach well over 2 m with its muzzle. Reconstructed shoulder height: 1.43 m.

Last but not least, the first true ruminants made their appearance in Europe at the end of the Eocene, following the same evolutionary radiation that led to such cursorial selenodonts as the xiphodontids and amphimerycids. Differing from such other artiodactyls as pigs and hippos, ruminants have a chambered stomach, which uses bacteria and protozoa to break down cellulose. Due to the extreme improbability that a soft-bodied feature such as a stomach would fossilize, recognition of ruminants in the fossil record is based mainly on some limb specializations that are shared by today's representatives, such as the fusion of the tarsal cuboid and navicular bones to form the cubonavicular bone. The most archaic ruminants, like the chevrotains that today inhabit the dense forests of Africa and eastern Asia (genera *Hyaemoschus* and *Tragulus*), bore a primitive limb configuration with still functional lateral digits and unfused central metacarpals. However, the first ruminants of the family Gelocidae, which appeared during the late Eocene in Europe, achieved further cursorial adaptations. Thus the limb became almost

FIGURE 2.15 *Reconstructed life appearance of* Paleotherium magnum
Browsing from relatively high branches with its head raised and its long neck stretched, *P. magnum* would have appeared quite unlike modern tapirs and would have vaguely resembled an okapi. With its robust limbs, it would have walked somewhat slowly and deliberately through the forests and woodlands of the late Eocene.

bidactyl, after the extreme reduction of the lateral segments, and the two central remaining metapodials were elongated, as in other advanced late Eocene selenodonts. These central metapodials finally fused into a single functional segment, forming the precursor of the ruminant "cannon bone." A similar condition exists today in some extant tropical forms like the moschoids of the genera *Hydropotes* and *Moschus,* which show a similar fusion of central metapodials, together with a reduction of lateral metapodials, but without the cranial appendages that are characteristic of most ruminants. *Gelocus,* from the late Eocene and the Oligocene in Europe, can be regarded as a feasible ancestor for this group of archaic ruminants.

The European terrestrial ecosystems at the end of the Eocene were quite different from those inherited from the Paleocene, which were dominated by archaic, unspecialized groups. In contrast, a diversified fauna of specialized small and large browsing herbivores—like the theridomorph rodents, the paleothere perissodactyls, and a wide array of bunodont and selenodont artiodactyls—characterized the late Eocene. From our perspective, they looked much more "modern" than those of the early and early-middle Eocene and perfectly adapted to the new late Eocene environmental conditions characterized by the spread of more open habitats. Nothing has been uncovered that would predict a sudden decline among such successful groups as the theridomorphs, paleotheres, and bunodont and selenodont artiodactyls. However, this is exactly what happened at the end of the Eocene, about 37 million years ago. The perissodactyls diminished in importance and variety, and most of the numerous late Eocene paleothere species died out before the end of that epoch. The theridomorph rodents also diminished in diversity to five or six genera, and several creodonts disappeared as well, such as a number of hyaenodontids and all the oxyaenids. However, the Asian immigrants *Pterodon* and *Hyaenodon* persisted. In any case, this decline in the diversity of the late Eocene terrestrial faunas was just an anticipation of the big breakdown that was going to occur at the beginning of the Oligocene.

CHAPTER 3
The Oligocene: A Time of Change

TODAY, ANTARCTICA IS AN ISOLATED CONTINENT, SEPARATED from Australia and South America by thousands of kilometers of ocean. Consequently, the shallow Antarctic Circumpolar Current flows continuously around the continent, preventing the warmer waters of the South Atlantic and Indian Oceans from reaching its coasts. But this was not the case during the Eocene, when Australia and South America were still attached to Antarctica, as the last remnants of the ancient Gondwanan supercontinent. Today's circumpolar current did not yet exist, and the equatorial South Atlantic and South Pacific waters went closer to the Antarctic coasts, thus transporting heat from the low latitudes to the high southern latitudes. However, this changed during the late Eocene, when a rifting process began to separate Australia from Antarctica. At the beginning of the Oligocene, between 34 and 33 million years ago, the spread between the two continents was large enough to allow a first phase of circumpolar circulation, which restricted the thermal exchange between the low-latitude equatorial waters and the Antarctic waters. A sudden and massive cooling took place, and mean global temperatures fell by about 5°C. In extensive parts of Antarctica, the snow that fell in the winter did not completely melt by the beginning of the next cold season, and as a consequence, a new glaciation began, more than 200 million years after the previous glacial phase during the Permian period. During a few hundred thousand years (the estimated duration of this early Oligocene glacial episode), the ice sheets expanded and covered extensive areas of Antarctica, particularly in its western regions. The presence of iceberg-rafted debris in the Indian Ocean sector of the Southern Ocean provides evidence of the spread of this glacial episode.

"LA GRAND COUPURE" AND THE END OF THE EUROPEAN ARCHIPELAGO

The onset of Antarctic glaciation and the growing of the ice sheets in western Antarctica provoked an important global sea-level lowering of about 30 m. Several shallow epicontinental seas became continental areas, including those that surrounded the European Archipelago. The Turgai Strait, which during millions of years had isolated the European lands from Asia, vanished and opened a migration pathway for Asian and American mammals to the west. Continentalization also increased in western Europe. To the western end of the Pyrenees, the Ebro Basin became fully continental, and the connection of the Iberian plate to the European mainland was completed. During the Oligocene and most of the Miocene, this basin was filled with fluviatile and lacustrine deposits that came from the surrounding reliefs of the Pyrenees and the emerging Iberian Range to the east. In north-central Europe, the Rhine graben progressed and opened a narrow connection between the North Sea and the Alpine Foreland Basin. The extension of this rifting process to the south through the Rhône graben had significant consequences for the configuration of the western Mediterranean, since part of the land to the east of Iberia and south of France shifted eastward, forming the future Corsica and Sardinia block.

The tectonic movements led to the final split of the Tethys Sea into two main seas, the Mediterranean Sea to the south and the Parathethys Sea, the latter covering the formerly open ocean areas of central and eastern Europe. While the Mediterranean Sea maintained a connection to the Atlantic Ocean, the Parathethys remained largely an intracontinental, enclosed domain, with reduced salinity and intermittent connections to the Mediterranean and the Indian Ocean. From the western Alps to the Transcaspian regions, thick deposits of dark laminated marls and shales were deposited in staving basins with reduced circulation and anaerobic bottom conditions.

After the retreat of the Turgai Strait and the emergence of the Parathethys province, the European Archipelago ceased to exist, and Europe approached its present configuration. The ancient barriers that had prevented Asian faunas from settling in this continental area no longer existed, and a wave of new immigrants entered from the east. This coincided with the trend toward more temperate conditions and the spread of open environments initiated during the late Eocene. Consequently, most of the species that had characterized the middle and late Eocene declined or became completely extinct, replaced by herds of Asian newcomers. The great Swiss paleontologist Hans Stehlin recognized this strong mammalian turnover in 1910, coining the term "La Grand Coupure" to designate it.

Among the first victims of the "Grande Coupure" were the paleotheres, which suffered a drastic reduction in their once-high diversity
levels. In fact, only a few species of the genera *Paleotherium* and *Plagiolophus* survived after the Eocene–Oligocene transition. The "Grande
Coupure" was even harder on the primates, which became extinct in
Europe at the beginning of the Oligocene (only some specimens of
Pseudoloris have been recovered in Spain at the very beginning of the
Oligocene, disappearing shortly after). Among the insectivorous stock,
the early Oligocene marks the end of the few persistent lineages of
archaic placentals like the leptictids and apatemyids, and likewise for
some rodent groups, such as the last ischyromyids of the genus *Plesiarctomys,* the ailuravids, and a number of brachydont theridomorphs.

The extinction of the arboreal primates and the reduction or extinction of several browsing groups like the paleotheres, ischyromyids,
and ailuravids are strong evidence for the retreat of the forests during
the early Oligocene and their replacement by open woodlands or even
drier biotopes.

In contrast, the didelphoid marsupials survived once again without
big losses. From nine lineages of the genera *Amphiperatherium* and *Peratherium,* five persisted and three new ones entered during the early
Oligocene. The creodonts continued their decline, and the hyaenodont
Pterodon as well as a number of species of *Hyaenodon* also became extinct.
However, new *Hyaenodon* species entered from Asia, maintaining the
presence of this group of archaic predators in Europe.

THE NEW OLIGOCENE UNGULATES

Among the most distinctive species to enter Europe after the "Grande
Coupure" were the first true rhinoceroses. From their first dispersal
during the early Oligocene, these perissodactyls had achieved a high
diversity and were going to characterize the mammalian faunas of Europe for millions of years, until the extinction of the last woolly rhinos
during the late Pleistocene. Among the first rhinos to enter the new
emerged European lands were the hornless hyracodonts and amynodonts and the horned diceratheres and menoceratheres.

The hyracodonts were hornless cursorial rhinos that evolved in
North America and Asia from the early Eocene *Hyrachius.* They were
represented in the European Oligocene by a sheep-size form, the still
little known *Egyssodon* (plate 4). Like most of the hyracodonts, *Egyssodon*
was a good runner, as indicated by its limb proportions. Besides these
small cursorial forms, another group of hyracodonts, the indricotherids,
attained gigantic dimensions in the Asian Oligocene. From the relatively
small, archaic *Forstercooperia,* the evolution of this group produced the
largest terrestrial mammals of any time. The giant *Paraceratherium* (also

known as indricothere or baluchithere) was 6 m tall at the shoulders and had a 1.5-m-long skull (figure 3.1). The males of this animal weighed around 15 tons, while the females were somewhat smaller, about 10 tons. Like the small hyracodonts, *Paraceratherium* lacked the characteristic horns of the rhinocerotids but bore large upper and lower incisors used to browse the leaves and branches at the tops of trees. At their westernmost dispersal, the indricotheres reached the Paratethyan shores (having been found in the Caucasus region, around Benara in Georgia).

Another peculiar group of rhino-related forms that also entered from Asia after the "Grande Coupure" were the amynodonts. These hornless, aquatic rhinos bore strong, tusklike incisors and had a short-legged, stout body. They probably resembled today's hippos in both aspect and lifestyle. The amynodonts first appeared during the middle Eocene in North America and reached maximal diversity during the late Eocene and early Oligocene, particularly in Asia. From this continent, members of the genus *Cadurcotherium* entered Europe after the "Grande Coupure," surviving there until the late Oligocene (although they persisted in Asia until the middle Miocene).

Hyracodonts, indricotheres, and amynodonts were all rhino-related groups that, surely, we could not easily identify as rhinoceroses if they were still alive. However, the early Oligocene in Europe also records the first entry of the true horned rhinos of the family Rhinocerotidae. Two rhinocerotid groups, the diceratheres and the menoceratheres, developed a pair of small horns on the tip of the nose independently. In the first diceratheres, the horns appeared as lateral protuberances at the sides of the nostrils. Close to the dicerathere stock was *Ronzotherium*, a slender, long-legged animal with tetradactyl forefeet and tridactyl hind feet. Humerus and femur were long, indicating good running capabilities. In contrast with the diceratheres, *Ronzotherium* was a hornless rhino that bore a pair of straight, elongated, tusklike lower incisors and two pairs of small upper incisors. A long diasteme separated these incisors from the remaining teeth. The rather retracted position of the nasal bones suggests that this animal bore a long upper lip, probably

FIGURE 3.1 *A scene from the Oligocene site of Benara, Georgia, with two creodonts of the species* Hyaenodon dubius *and a group of indricotherids of the genus* Paraceratherium

During the Oligocene, the large, wolf-size hyaenodontids like *Hyaenodon dubius* were the largest mammalian land predators not only in the Caucasus but in all of Eurasia, so, as this illustration makes clear, adult indricotheres were virtually immune to predation. The fragmentary remains recovered at Benara correspond to an indricotherid of rather moderate size, with a reconstructed shoulder height of about 4 m, but large specimens from Mongolia could attain heights of more than 5 m.

used, as by modern rhinos, to browse high foliage. The oldest representatives of *Ronzotherium* in Europe were rather small, but this genus demonstrated a trend toward increasing size, and the last species reached large dimensions.

Much less common in early Oligocene localities, the menoceratheres included *Epiaceratherium,* a small horned rhino. This form retained a rather complete dentition in which the canines were still present. They also bore elongated, lanceolate lower incisors that curved upward and were longer in the male than in the female. Although scarcely represented in few early Oligocene localities, *Epiaceratherium* is important because it was the first European link of one of the most successful groups of Oligocene and Miocene rhinos: the aceratheres. Characteristically, most aceratherine rhinos showed, like *Epiaceratherium,* a strong sexual dimorphism affecting the incisors.

Besides rhinos, a second group of Oligocene perissodactyls that entered Europe after the "Grande Coupure" were the chalicotherids. This group of bizarre large browsers prolonged the trend observed in several Eocene perissodactyls to enlarge the forelimbs against the hind limbs, a typical adaptation to reach the higher branches of trees by standing up over the hind feet. This trend reached an extreme in some Miocene representatives that developed gorilla-like limb proportions, with very long forearms and short hind limbs. Moreover, the chalicotheres bore claws at the ends of their arms instead of the typical hoofs of the odd-toed ungulates. These claws enabled them to grasp small branches and leaves with the help of their forefeet. This group of bizarre ungulates evolved from the lophiodonts, and some of these Eocene tapirlike perissodactyls are now recognized as true chalicotheres (*Lophiaspis*). In the Oligocene, the first chalicotheres entering Europe after the "Grande Coupure" belonged to the genus *Schizotherium.* Like other members of the subfamily Schizotherinae, and in contrast to the more common middle and late Miocene chalicotheres, *Schizotherium* had forelimbs that were still capable of conventional locomotion over the ground. Unlike the Miocene chalicotheres, *Schizotherium* bore a specialized, semihypsodont dentition.

While the "Grande Coupure" completely changed the perissodactyl representation from the late Eocene paleothere-dominated faunas to the early Oligocene rhino-dominated faunas, the effects on the artiodactyls were more limited. Several families and genera that were already present in the late Eocene pursued their evolution into the early Oligocene. Among the archaic dichobunids, several forms like *Dichobune, Tapirulus,* and *Diplobune* persisted into the Oligocene. In contrast, a number of late Eocene families among the bunoselenodont forms, like the dacrytherids, xiphodontids, and anoplotherids, became extinct. This may have been a result of competition from the more sophisticated gelocid ruminants, which in the early Oligocene pursued diversification into several genera, like *Gelocus, Lophiomeryx,* and *Bachitherium* (figures 3.2 and 3.3). However, the spread of these moschoid ruminants seems not to have affected the small selenodont cainotherids, which not only survived the "Grande Coupure" but even became common among all the Oligocene faunas. These cainotherids included two genera of different dimensions: the rabbit-size *Plesiomeryx* and the larger *Caenomeryx* (plate 3).

Among the bunodont artiodactyls, the anthracotheres were the most common, persisting in a variety of genera such as *Anthracotherium, Bothriodon,* and *Elomeryx.* Some of them, like *Anthracotherium,* were very large forms (the size of a hippo), with powerful jaws, big canines, and shovel-like incisors. Others, like *Bothriodon* and *Elomeryx,* were smaller, pig-size forms with slenderer limbs and elongated skulls. The canines were, in general, more reduced than in the large *Anthracotherium* but were long

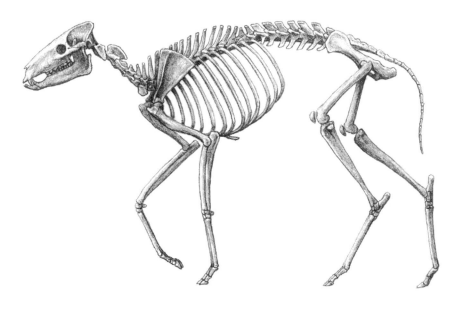

FIGURE 3.2 *Skeleton of the ruminant* Bachitherium insigne

A beautifully preserved, articulated skeleton of this species was found at the site of Cereste, France, giving a clear picture of the animal's body proportions. *Bachitherium*'s hind limbs were considerably longer than the forelimbs, the head was rather long, and the neck was short. The tusklike teeth in the mandible were not canines, as one would expect, but were actually caninelike first premolars—which is why the upper tusks are in front of the lower ones, contrary to the norm among mammals. Reconstructed shoulder height: 54 cm.

FIGURE 3.3 *Reconstructed life appearance of* Bachitherium insigne

In life, *Bachitherium* would have looked rather odd to a modern observer, due to the combination of the antelope-like, elongated limbs and the short neck and massive head. The dentition indicates a browsing diet, and the animal's limb proportions, like those of many woodland ruminants, would have allowed it to move easily among the shrubs, eluding predators with powerful leaps and sudden, zigzagging turns.

and posteriorly serrated in the males of *Elomeryx*—a clear case of sexual dimorphism (plate 3).

Despite the diversity and abundance of the Oligocene anthracotheres, new Asian immigrants joined them in the early Oligocene terrestrial ecosystems. Among them, the most impressive were the gigantic entelodontids, a rather common group during the Oligocene in Europe and North America. The entelodonts, like the American *Archaeotherium* and the European *Entelodon*, were large, pig-related forms that attained gigantic sizes for a suiform, with large skulls around 1 m long (figures 3.4 and 3.5). They had massive bodies, with thoracic vertebrae displaying long neural apophysis and ending in a short tail. A short and robust neck held a large skull that was broad and low, with enlarged zygomatic arches. The dentition was complete and resembled that of some archaic

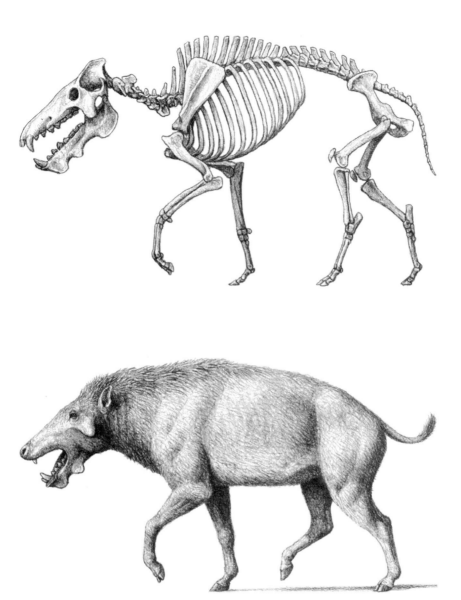

FIGURE 3.4 *Skeleton and reconstructed life appearance of the entelodontid* Entelodon deguilhemi

The anatomy of European entelodons was almost completely unknown until the finding of abundant cranial and postcranial remains of *Entelodon deguilhemi* at the site of Villebramar, France. This material shows the animal to have been quite large, with a height of 1.35 m at the shoulders and a skull 65 cm long. Broadly piglike in appearance, entelodons differed from true suids in having a completely fused radius and ulna, as well as the tibia and fibula, and in having lost the lateral digits, thus being didactyl. These features, as well as the greater elongation of the metapodials, point to a moderate adaptation to cursorial locomotion in relatively open terrain.

condylarths, like the arctocyonids and mesonychids, rather than that of an artiodactyl. The entelodonts bore simple, conical premolars that looked rather "reptilian" in shape. The forelimbs were more robust and stronger than the hind limbs, and the distal segments of the limbs, like the tibia, were certainly shorter than the proximal ones, like the femur. This indicates a low capacity for a fast cursorial locomotion. However— and this points to a difference from the hippolike anthracotheres—the entelodonts had only two digits, having already lost the lateral ones. Their conical dentition and lateral tooth wear suggest that they were mainly carrion-eaters with a good capability for bone-crushing. The species *Entelodon magnum* attained a wide distribution in Europe, from Spain, France, and Germany to Romania and the Caucasus.

Not so impressive as the big entelodontids, but even more significant, were the first true pig-related forms of the family Paleochoeridae to disperse from Asia, like *Palaeochoerus* and *Doliochoerus*. Unlike the five-toed, selenodont anthracotheres, these first suoids were small pigs with tetradactyl limbs (the forelimbs had already lost the first digit) and a complete dentition, including well-developed incisors, vertical canines (not so big as in some anthracotheres), single premolars, and low-crowned, bunodont molars. There was a trend for the molars to develop several accessory cusps, as a response to a fully omnivorous diet.

THE NEW OLIGOCENE SMALL-MAMMAL COMMUNITIES

One of the most significant features of the early Oligocene small-mammal communities was the first entry of lagomorphs into Europe. The lagomorphs—that is, the order of mammals that includes today's hares and rabbits—originated very early on the Asian continent and from there colonized North America. The presence of the Turgai Strait prevented this group from entering Europe during the Eocene. After

FIGURE 3.5 *Skull, musculature, and reconstructed head of* Entelodon deguilhemi
A nearly complete skull and mandibles from Villebramar show the clear resemblances between the skull of *Entelodon deguilhemi* and species of the American genus *Archaeotherium*. Both animals had elongated heads with flaring zygomatic arches and mandibles with bizarre, knobby ventral projections. The coronoid process in the mandible was low, resembling that of saber-toothed cats and implying an adaptation to increased gape, and the wear in the tusklike canines shows that they were used for biting large objects, not only for display and occasional intraspecific fighting, as in pigs and hippos. These features, together with the shape and type of wear in the cheek-teeth, suggest that these animals were omnivores and scavengers and occasionally may have been active predators.

the retreat of this shallow sea, the ochotonids of the genus *Shamolagus* dispersed into Europe during the early Oligocene. The ochotonids are a family of rabbit-size lagomorphs whose members (known as pikas) settled on the steppes of western and Central Asia. In contrast to today's hares and rabbits, the pikas (*Ochotona*) bear shortened ears.

Among the insectivores, after the demise of the archaic placental groups, the modern talps (*Eotalpa*), shrews (*Quercysorex*), and hedgehogs (*Tetracus*) became dominant.

Among the rodents, the most common were still the theridomyids, which declined in diversity and were represented mainly by lophodont forms like *Pseudoltinomys, Theridomys, Blainvillimys,* and *Taeniodus* (figure 3.6). Common among them was the trend to develop increasingly hypsodont cheek-teeth, which reached an extreme in *Archaeomys* and *Issiodoromys*. Among glirids, the other rodent family that originated in the middle Eocene, *Gliravus* persisted in the Oligocene and even split into several species of different sizes and morphologies. However, the main difference between the Eocene and Oligocene rodent faunas in Europe was the sudden appearance after the "Grande Coupure" of several new families representing the modern groups that inhabit this continent, such as hamsters, squirrels, and beavers, as well as extinct members of this order.

The first squirrels of the family Sciuridae to enter Europe are usually assigned to the genus *Paleosciurus*. Surprisingly, despite some small differences in their dentition, they probably looked very much like today's squirrels. This is an extraordinary case of morphological stability through time. Various features, particularly the limb proportions, suggest that the first *Paleosciurus* species (*P. goti*) were ground squirrels. As in modern sciurids, the tibia and radius were relatively short with respect to the femur and humerus.

Other sciuromorph rodents that entered Europe after the "Grande Coupure" were the beavers of the genus *Steneofiber*. Beavers bear a characteristic pentalophodont molar pattern, like that of the theridomyids, but their origin lies close to that of the squirrel group. The first castorids of the genus *Steneofiber* were about half the size of today's *Castor* but probably already had an aquatic lifestyle close to that of present-day water voles.

A third family of extant rodents that entered Europe for the first time during the early Oligocene was the aplodontids. This group has today only one species, *Aplodontia rufa,* the mountain beaver of North America. This is a large marmotlike rodent that excavates burrows in the mountains of Oregon. In the Oligocene, the Aplodontidae was a successful family represented by a number of genera, such as *Sciurodon* and *Plesispermophilus.* The aplodontids were browsing rodents that

FIGURE 3.6 *Skeleton and reconstructed life appearance of the theridomorph rodent* Pseudoltinomys gaillardi

A remarkably complete, articulated skeleton of this rat-size rodent was discovered
at the site of Cereste, preserving such uncommon details as the impression of hairs
in the long, bushy tail. The body proportions of *Pseudoltinomys gaillardi*, with short
arms and long hind limbs well adapted to jumping, indicate a hopping type of
locomotion like that of modern gerbils and a similar adaptation to life in semiarid
environments. Reconstructed head and body length: 21 cm.

showed a trend toward developing selenodont upper molars. Moreover,
new connections appeared between the crests, while the cusps diminished in size and were more and more transformed into ridges. This
was accompanied by a size increase in the premolars compared with the
molars. Altogether, their dentition suggests a strong adaptation to life
in the forest and a diet based on leaves.

Another highly significant group of rodents that appeared in Europe
during the early Oligocene was the eomyids, today completely extinct,
but one of the most characteristic families during most of the Oligocene
and Miocene. The eomyids' dentition was composed of one premolar
and three molars in each jaw. All the eomyids were characteristically
small, about 5 cm long, usually among the smallest rodents of their time.

Eomys, the first member of this family in Europe, bore brachydont, low-crowned molars with distinctive cusps united by ridges. Remarkably, *Eomys* had skin folds connecting its front and rear limbs as a patagium, which enabled it to glide from tree to tree like today's flying squirrels.

But the most characteristic immigrants during the early Oligocene were the cricetids of the genus *Atavocricetodon.* The cricetids are today represented in Europe by hamsters, reduced to three or four species (*Cricetus cricetus, Mesocricetus auratus, Cricetulus migratorius, Tcherskia triton*). These cricetids are typical inhabitants of the cold steppes of eastern Europe and Central Asia, and their limited representation in today's European ecosystems does not reflect their importance in the history of the Cenozoic mammalian faunas of Eurasia. After its first entry following the "Grande Coupure," this group experienced extraordinary success, diversifying into several genera and species. Even more significantly, the cricetids gave rise to the rodent groups that were going to be dominant during the Pliocene and Pleistocene—that is, the murids (the family of mice and rats) and arvicolids (the family of voles). The first cricetids in Europe (*Atavocricetodon, Pseudocricetodon*) were small rodents whose tooth morphology resembled that of the coeval eomyids— that is, brachydont molars with distinct, bunodont cusps, connected by low crests. However, in cricetids (as in their murid and arvicolid descendants) each tooth row was composed of only three molars, having lost the last premolar. The first molar of the cricetids was an enlarged tooth that differed in size and shape from the eomyids' small first lower molar and was comparable to the deciduous molar that the members of the latter family still show in young individuals. It is, therefore, highly probable that the enlarged first molar of the cricetids originated from the retention of this deciduous tooth in the adult, a process known in evolutionary biology as neoteny (that is, the retention of juvenile traits in the adult).

THE NEW OLIGOCENE CARNIVORES

The "Grande Coupure" also witnessed a strong faunal turnover among the true carnivores, which from this time on joined the hyaenodont creodonts in the large-predator guild. The arctoids of the genus *Cynodictis* disappeared in favor of forms like *Amphicynodon,* which approached the bear group in anatomy (figures 3.7 and 3.8). The members of this genus varied in size from a genet to a raccoon, with a large and low skull and big canines. They differed from *Cynodictis* in their more-bunodont, less-puncturing dentition, which indicates a more omnivorous diet. This trend toward omnivorism was even more pronounced in the closely related *Pachycynodon,* which bore molars with low, blunt cusps better adapted to grind than to shear.

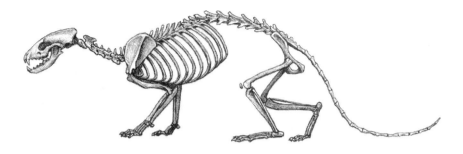

FIGURE 3.7 *Skeleton of the early caniform carnivore* Amphicynodon leptorhynchus
This small carnivore, no larger than a house cat, is well known thanks to cranial and postcranial fossils from Quercy, France. It had a generalized skeleton with long, flexible back and plantigrade or semiplantigrade feet capable of considerable lateral rotation. These features suggest that the animal probably would have foraged on the ground for at least part of the time, but could readily have climbed trees either in pursuit of prey or to escape larger, terrestrial predators like the hyaenodontid creodonts. Reconstructed shoulder height: 18 cm.

In addition, new carnivore families, like the nimravids, appeared, approaching today's catlike carnivores in their more advanced dental and locomotor adaptations. The nimravids were once regarded as true felids (the family that includes today's big and small cats) because of their similar dental and cranial adaptations. For instance, they showed a lengthening of the cutting blade in the carnassials at the expense of the grinding area. Also, the postcarnassial molars were often reduced or even lost. But one of the more distinctive attributes of the nimravids was their long, laterally flattened upper canines, which were similar to those of the Miocene and Pliocene saber-toothed cats like *Machairodus* and *Megantereon*. However, most of these features have proved to be the result of a similar adaptation to hypercarnivorism, and the nimravids are now placed in a separate family of early carnivores whose evolution paralleled that of the large saber-toothed felids. As large hypercarnivorous predators, they had a diet based only on meat, having lost the ability to ingest large amounts of plant matter. The first early Oligocene representatives of this family were of the genus *Eusmilus*, characterized by advanced hypercarnivorous adaptations (figures 3.9 and 3.10). This dirk-toothed cat was the size of a lynx but bore enormously enlarged upper canines, which when the mouth closed were located in a sort of sheath in the mandible. Also common during the early and middle Oligocene was *Nimravus* (plate 4). This nimravid differed from *Eusmilus* in its less-developed upper canines, which approached the size of those in today's felids. Another nimravid, *Quercylurus,* was the largest carnivore

of its time, reaching the dimensions of a modern brown bear. It was a massive, plantigrade form whose robust limbs resembled more those of a bear than a cat.

Despite the presence of large, advanced nimravids, the early Oligocene also records a number of small forms similar to the true cats. These were archaic feloids that resembled modern civets and genets (family Viverridae) in size and morphology. The most common genus in the early Oligocene was *Stenoplesictis*. In contrast to *Cynodictis* and the coeval *Amphicynodon*, the members of *Stenoplesictis* had higher carnassials with better differentiated blades. They also exhibited a trend toward developing a flattened dorsal face and reducing the second molar (the third one was already absent). In many respects, *Stenoplesictis* and the closely related *Palaeoprinodon* were close to today's palm civet(*Nandinia binotata*), an arboreal viverrid of about 3 kg that inhabits the tropical forests of central Africa (Peigné and Bonis 1999a) (figure 3.11).

THE LATE OLIGOCENE

About 30 million years ago, a new glacial phase began, and for 4 million years Antarctica was subjected to multiple glaciation episodes. The global sea level experienced the largest lowering in the whole Cenozoic, dropping by about 150 m (Haq et al. 1987). A possible explanation for this new glacial event lies in the final opening of the Drake Passage between Antarctica and South America, which led to the completion of a fully circumpolar circulation and impeded any heat exchange between Antarctic waters and the warmer equatorial waters. A second, perhaps complementary cause for this glacial pulse is probably related to the final opening of the seaway between Greenland and Norway. The cold Arctic waters, largely isolated since the Mesozoic, spread at this time into the North Atlantic. The main effect of this cooling was a new extension of the dry landscapes on the European and western Asian lands. For instance, we know from pollen evidence that a desert vegetation was dominant in the Levant during the late Oligocene and earliest Miocene (Goldsmith et al. 1988).

This glacial event led to the extinction of several forms that had persisted from the Eocene, such as the last paleotheres of the genus

....

FIGURE 3.8 *A scene from the site of Itardies, France, with* Amphicynodon leptorhynchus *attacking two rodents of the genus* Pseudoltinomys

The body proportions of *Amphicynodon* suggest that it would have hunted small vertebrates with a stalk-and-pounce technique, not unlike that of modern procyonids of the genus *Bassariscus*, the so-called ringtail cats. Such a hunting technique would have been adequate for taking such agile saltatorial rodents as *Pseudoltinomys*.

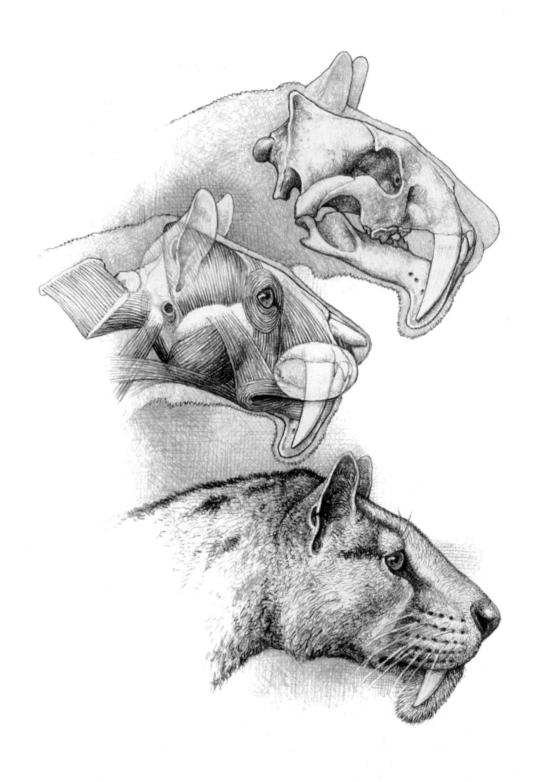

Plagiolophus, the large *Entelodon,* and most of the anthracotheres. The theridomyid rodents also declined, and only two hypsodont genera, *Issiodoromys* and *Blainvillimys,* survived. Remarkably, *Issiodoromys* developed hypsodont cheek-teeth that basically became two columnar prisms. This kind of dental evolution strongly resembles that of some South American caviomorph rodents that today inhabit the steppe landscapes of the pampas. Like them, *Issiodoromys* had inflated auditory bullae, a common feature in rodents living in desert environments. At the end of the Oligocene, *Issiodoromys* was the only survivor of the once highly diversified theridomorphs.

In contrast, the glirids not only survived in the new hard conditions but even diversified into new genera and species. From the conservative *Eliomys*-like molar design of *Gliravus,* based on few crests, new genera developed multicrested teeth with several parallel ridges. Some of them, like *Bransatoglis,* were larger than *Gliravus* and had a molar pattern close to that of the Edible dormouse (*Glis*). Others, like *Peridyromys* and *Microdyromys,* were smaller glirids with dental patterns similar to those of today's Balkan dormice (*Dryomys* and *Myomimus*). The persistence of the glirids in the new regime of seasonality was probably related to their hibernating capabilities, which enabled them to survive the harsh conditions of the dry season.

The small, gliding eomyids also expanded at this time, the genus *Eomys* diversifying into a number of species and becoming common in the late Oligocene record. From the bunodont and brachydont pattern of the first species of this genus, one lineage tended to develop increasingly lophodont molars in which the ridges were higher and more important than the cusps. In the more advanced species of this lineage, like those of the genus *Rhodanomys,* a fully lophodont pattern without recognizable cusps was already extant.

....

FIGURE 3.9 *Skull, musculature, and reconstructed head of the nimravid* Eusmilus bidentatus

The partial skulls and mandibles recovered in the nineteenth century from the site complex of Quercy remain the best fossils of this species known to this date, allowing a detailed reconstruction of the head of this highly specialized dirk-toothed nimravid. Although *Eusmilus bidentatus* was no larger than a modern lynx, the adaptations for gape seen on its skull and mandible are more advanced than in any of the felid sabertooths of the European Pliocene and Pleistocene. The low coronoid process, elevated occiput, lowered glenoid fossa, enlarged and ventrally projected mastoid, and atrophied paroccipital process were all adaptations for biting with the enormously enlarged upper canines. Such a bite required the pull of the neck muscles to help in depressing the head and sinking the canines into the flesh of the prey animal, a complex set of muscular actions called the canine shear-bite. Basal length of skull: 13 cm.

However, the rodent group that experienced a higher diversification and success under the new conditions of the late Oligocene were the cricetids, with several genera of different sizes and dental morphologies, such as *Heterocricetodon, Paracricetodon, Cincamyarion, Eucricetodon,* and *Adelomyarion.* All these genera had complicated cheek-teeth in which a number of ridges connected the cusps. At the extreme was *Melissiodon,* a large cricetid whose molar cusps were transformed into ridges with new connections between them. As in some browsing flying squirrels that bear a similar dental pattern, *Melissiodon* was probably an arboreal hamster with a diet composed basically of leaves. While *Melissiodon* persisted up to the early Miocene, the other group of browsing rodents, the selenodont aplodontids, persisted in central Europe until the middle Miocene.

Among the carnivores, the late Oligocene saw the decline and local extinction of the large nimravids. In contrast, the group of archaic feloids that had arisen during the early Oligocene with *Stenoplesictis* and *Palaeoprinodon* continued its evolution into the late Oligocene and diversified into a number of genera like *Haplogale, Stenogale,* and *Proailurus.* Most of them, like *Stenogale,* showed characteristics similar to those of modern felids, such as a trenchant dentition and a short snout, with the reduction of premolars and the loss of molars. In particular, *Proailurus* can be regarded as a true felid and, plausibly, as the ancestor of today's cats. This ocelot-size carnivore already displayed the essential features characteristic of the felids, such as a shortened face and a reduced cheek-tooth row. Remarkably, it resembled the modern fossa, a catlike viverrid from Madagascar, with short, robust semiplantigrade limbs and retractile claws that enabled it to climb and jump in the trees in search of small arboreal vertebrates. *Proailurus* pursued its evolution during the early Miocene, leading to the first representatives of the genus *Pseudaelurus* (figure 4.1).

The other group of large carnivores that spread during the late Oligocene were the "bear-dog" amphicyonids, which from that time on became quite diverse, with many different ecological adaptations. The members of this family bore elongated skulls, with well-developed sag-

FIGURE 3.10 *A scene from the Oligocene site of Villebramar, France, with two* Eusmilus *leaving their prey to two entelodons of the species* Entelodon deguilhemi

By virtue of their huge size, it is likely that entelodons were able to displace any European Oligocene predator from its rightful kill. Despite its fearsome sabers, *Eusmilus* did not stand a chance of defending its kill against the cow-size scavengers. The environment shown in this reconstruction corresponds to the dry woodland that was becoming widespread in Europe at the time, allowing the open-terrain-associated faunas from Asia to extend farther into the west.

ittal crests supporting powerful mandibular muscles. Their dentition was similar to that of the canids in shape and morphology, characterized by enlarged molars and a tendency to develop reduced premolars and carnassials. Enlarging the molars means broadening the grinding surface and was probably an adaptation to crush meat and bones, in contrast to the shearing abilities of other carnivore families, such as the nimravids and felids. Despite their variety of feeding adaptations, the amphicyonids shared a rather conservative postcranial anatomy. The strong fusion seen in the sacral vertebrae probably served to support the muscles controlling a long tail, which in the Miocene *Amphicyon longiramus* was composed of twenty-eight vertebrae (Olsen 1960). Such long tails are not rare among the big cats and are usually used for balancing while running. However, despite some Miocene exceptions, the amphicyonids were not cursorial predators but generalized mesocarnivores that preyed on small or medium-size herbivores by stalking and pouncing (Viranta 1996). As mesocarnivores, they were able to switch to an herbivorous diet, despite their preference for meat. Also, because of the shape of their ungual phalanges, very curved in the case of *Daphoenus vetus,* the amphicyonids might have had some climbing abilities (Olsen 1960). Despite their large size, climbing abilities would not be surprising, given that large ursids like the modern black bear are agile climbers. The first Oligocene amphicyonids like *Pseudocyonopsis* and *Cynelos* were relatively small unspecialized mesocarnivores preying on small vertebrates. Their dentition resembled that of today's canids, showing both shearing and grinding function in roughly equal proportions. They bore well-developed premolars and a lower carnassial with a long blade in the trigonid. In contrast, a third amphicyonid form, *Ysengrinia,* developed an opposite trend toward hypercarnivorism, as reflected by the strong reduction of its molars.

The late Oligocene saw, in addition to the bearlike amphicyonids, the spread of the first true ursids, represented by *Cephalogale*. The members of this genus did not have the massive body dimensions of today's bears but were medium-size omnivores in which the crushing features dominated the dentition, which maintained some slicing characteristics. The postcarnassial teeth were considerably enlarged, while the grinding part of the carnassial (known as talonid) was broad and made up a large part of the tooth. *Cephalogale* probably arose from an arctoid form close to the early Oligocene *Amphicynodon*.

Another group of carnivores that spread successfully during the late Oligocene were the mustelids, the family that includes today's martens, badgers, skunks, and otters. Although none of the present-day mustelids could yet be recognized in the late Oligocene, there was at this time a first branching of a mesocarnivore and a hypercarnivore stock. The members of *Amphictis, Bavarictis,* and *Pseudobassaris* were rather small,

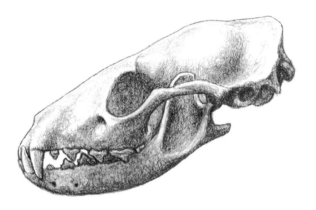

FIGURE 3.11 *Skull and reconstructed head of* Palaeoprionodon lalandi

Beautifully preserved fossils of this species from Quercy reveal a skull similar to that of a modern genet, suggesting a similar diet based largely on small vertebrates.

stout forms with thickened molars. In contrast, other genera like *Mustelictis, Plesictis, Plesiogale,* and *Paragale* consisted of larger species with more shearing dentition (plate 3). This Oligocene branching was in the same direction as the first evolutionary radiation of modern mustelids, which would occur during the Miocene.

In contrast to these successes, the creodonts of the genus *Hyaenodon,* which had survived all periods of crisis since the Eocene, declined during the late Oligocene. The last *Hyaenodon* in Europe was recorded at the end of the Oligocene in the locality of Coderet, France, and did not survive into the Miocene. This was the end in Europe of a long-lived group of successful carnivorans that had filled the large-predator guild for millions of years. However, as with other Oligocene groups, like the anthracotheres, the hyaenodonts persisted in Africa and, from there, made a short incursion into Europe during the early Miocene (chapter 4).

THE LATE OLIGOCENE UNGULATES

During the late Oligocene, the early Oligocene representation of hyracodont rhinos disappeared from Europe. Similarly, the once widely expanded rhinos of the genus *Ronzotherium* disappeared without descendants. In their turn, the rare menoceratheres and their offshoots became the dominant rhinos. From the early Oligocene *Epiaceratherium,* two genera, *Menoceras* and *Protaceratherium,* expanded across Europe, persisting there until the early Miocene. *Menoceras* was a small-horned rhino that retained its fourth digit in the forefoot as an archaic feature. *Protaceratherium* was a hornless rhino with slender limbs and good running capabilities. Both forms bore low-crowned cheek-teeth adapted to browsing leaves or brush. From this rhinocerotid stem, rhinos diverged during the late Oligocene into three major groups: the rhinocerotines, the aceratherines, and the teleoceratines. The three followed very different evolutionary pathways throughout the Miocene: the rhinocerotines bore a pair of horns and reduced incisors; the aceratherines were browsing hornless rhinos with a prehensile lip; and the teleoceratines became hippolike, single-horned animals.

The aceratheres bore strong, tusklike lower incisors that they probably used in conjunction with their well-developed prehensile lips, becoming the most conspicuous rhinocerotids during the Miocene. The first recognized member of this group was the late Oligocene *Mesaceratherium,* characterized by its narrow skull and slender limbs. Like other members of the group, *Mesaceratherium* bore strong, upwardly curved incisors, supported by a robust mandible with a broad mandibular symphysis. They operated against other large, chisel-shaped upper incisors.

Unlike the aceratheres, the teleoceratheres never lost their first upper pair of incisors, which retained their shearing function. The teleoceratheres were massive, short-legged rhinos in which the distal segments of the limbs were considerably shortened. The first late Oligocene representative of this group was *Brachydiceratherium* (*B. lemanense*), a large rhino that still had a tetradactyl forefoot. The skull was long and narrow and bore a small horn at the tips. The lower incisors were strongly curved and supported by a rather broad mandibular symphysis, and the molars were broad and low-crowned. All these features indicate that *Brachydiceratherium* probably had a lifestyle close to that of today's tapirs, passing most of its time in the shallow waters of lakes and rivers and eating the soft vegetation of the margins.

Rhinocerotines, aceratherines, and teleocerathines illustrate quite well the different use of horns and incisors as potential weapons. In the early rhinocerotids, such as *Menoceras,* the possession of horns was complemented with the retention of well-developed incisors. However, the later evolution of the group strengthened these two weapons in different ways. Among the rhinocerotines, the strengthened weapons were the horns, while the incisors remained small or even disappeared. To use the horns, the head was lowered, in a position characteristic of that of today's black and white rhinos. In contrast, among aceratheres and teleoceratheres, the incisors were strengthened, becoming actual tusks, while the horns were lost; the head maintained a horizontal position, and the limbs tended to be short to permit browsing on low vegetation. This dichotomy is still seen among today's rhinocerotids. The Asiatic rhinos use their horns mainly in fights between males, while their principal defense is their large tusklike incisors. In contrast, in the African rhinos the horns are the most conspicuous weapon, and the head adopts a lowered position.

Another distinctive feature of the late Oligocene was the first appearance in Europe of the true tapirids of the genus *Protapirus.* This archaic tapir differed in a number of characteristics from today's members of the family, such as in its still archaic dentition with reduced, nonmolarized premolars. In contrast with those of the Eocene lophiodonts and other Eocene tapirlike forms, *Protapirus*'s skull anatomy and retracted nasals indicate that a proboscis like that of today's tapirs was already present at the end of the snout.

Among the artiodactyls, the diversification of the moschoid ruminants, direct descendants of the gelocids, characterized the late Oligocene. Genera like *Prodremotherium* and *Bedenomeryx* were close to the ancestral *Gelocus* but developed longer metapodials and changed some details of this segment of the limb anatomy, such as closing the distal end of the cannon bone.

The late Oligocene faunas, therefore, where very different from those of the late Eocene and were characterized by the quick spread of "modern" browsing perissodactyls, such as tapirs and the rhinocerathine, aceratherine, and teleoceratine rhinos. Similarly, the artiodactyl community was composed mainly of advanced cervoids with derived dental and locomotor features that indicate an adaptation to more open vegetation. The predator guild was at that time already composed of a variety of carnivores of different sizes and diets. While the hyaenodonts persisted in the late Oligocene, the large-predator guild was increasingly occupied by Oligocene newcomers like the amphicyonids. After the long crisis of the late Eocene and the early Oligocene, the European realm seemed to enter a period of climatic stability. But this was not going to occur, since new events were going to deeply affect the European terrestrial ecosystems.

CHAPTER 4
The Early to Middle Miocene:
When the Continents Collide

A FTER A GRADUAL WARMING DURING THE LATE OLIGOCENE, global temperatures reached a climatic optimum during the early Miocene (the warmest interval since the late Eocene; Flower and Kennett 1994). Shallow seas covered several nearshore areas in Europe, such as the Rhône and Aquitanian Basins in France and the Tagus Basin in Atlantic Iberia, as a consequence of a general sea-level rise. A broad connection was established between the Indian Ocean and both the Mediterranean and Paratethys Seas, which were populated by numerous warm-water Indo-Pacific immigrants. Widespread warm-water faunas including tropical fishes and nautiloids have been found, indicating conditions similar to those of the present-day Guinea Gulf, with mean surface-water temperatures around 25 to 27°C. Important reef formations bounded most of the shallow-water Mediterranean basins. In the southern hemisphere, the southward migration of tropical and subtropical marine fauna in the early Miocene also indicated this climatic trend. Warm-water, larger benthic foraminifers and mollusks reached their southernmost distribution at this time (Hornibrook 1992). Reef-building corals that today inhabit the Great Barrier Reef within a temperature range of 19 to 28°C became well established on North Island, New Zealand (Hornibrook 1992).

The early Miocene climate was warm and humid, indicating tropical conditions (figure 4.1). Rich, extensive woodlands with varied kinds of plants developed in different parts of southern Europe, such as the Lisbon Basin in Portugal (Antunes and Pais 1983) and the Vallès-Penedès Basin in northeastern Spain (Bessedik and Cabrera 1985). In the Vallès-Penedès Basin, an abundance of evergreen broad-leaved elements and megathermic taxa, similar to today's evergreen broad-leaved forest in

eastern Asia, characterized the forests (Bessedik 1984). This evergreen forest, known as Oak-Laurel Forest, or Laurisilva, occurs today at 200 to 300 m in altitude in this part of Asia and in some relict areas like the Canary Islands. It is composed primarily of cupuliferous (the oak group) and lauraceous trees (laurel, cinnamon, camphor trees, and *Phoebe*). *Podocarpus,* pines, hemlocks, magnolias, fig trees, and rattan palms are found at higher altitudes. The conditions were probably similar to those of the intertropical zone, where local topographic conditions allowed the persistence of megathermic elements (such as *Bombax, Grewia, Symplocos,* and *Avicennia*) and the development of coral reefs. The climatic optimum of the early Miocene also led to a maximum development of mangroves. These subtropical floras extended as far north as eastern Siberia and Kamchatka (Volkova et al. 1986). The occurrence of subtropical molluscan species (60%) and temperate flora (33% of beech) in boreal Kamchatka (Gladenkov 1992) further illustrates the extent of this early to middle Miocene warmth.

The Early Miocene Mammalian Faunas

The earliest Miocene mammalian faunas were basically composed of species persisting from the late Oligocene. Therefore, the earliest Miocene terrestrial ecosystems maintained a similar structure based on small to medium-size browsers (archaic artiodactyls, moschoids, and rhinos), with the amphicyonids as the main carnivores.

Among the perissodactyls, the early Miocene record indicates the presence of a number of small to medium-size tapirs that persisted from the late Oligocene (*Protapirus*). Two more-advanced tapirs, *Paratapirus* and *Eotapirus,* bore more-derived dentition with molarlike premolars.

....

FIGURE 4.1 *A scene from the early Miocene site of Saint-Gérand-le-Puy, France, with the early felid* Proailurus lemanensis

The extremely fossil-rich deposits of Saint-Gérand-le-Puy, France, accumulated along the shores of a lake surrounded by subtropical evergreen forests and woodlands. The excellent sample of *Proailurus* fossils from Saint-Gérand includes many complete limb bones, showing that this ocelot-size carnivore had short, robust limbs and was semiplantigrade, remarkably resembling in structure and locomotion the modern fossa, a catlike viverrid from Madagascar. In the upper right is a primitive squirrel of the genus *Paleosciurus,* just becoming aware of the presence of its potential predator. The rodent was as well adapted to life in the trees as modern squirrels, but the early cat was probably a better climber than most modern felids and a likely threat to squirrels. We must remember that the similarly adapted fossa is able to hunt sifakas and other lemurs, which are extraordinarily agile in the branches. Reconstructed shoulder height of *Proailurus:* 38 cm.

All these forms disappeared in the early Miocene, not reaching the middle Miocene. A similar picture can be drawn for the rhinoceroses, with some representatives—such as *Protaceratherium, Menoceras,* and *Mesaceratherium*—persisting from the late Oligocene (figure 4.2).

Among the artiodactyls, the peculiar nonruminant cainotheres persisted during all the early Miocene, represented by the genus *Cainotherium* (figure 4.3). These rabbit-size selenodonts are still common in several localities of western and central Europe, where they are represented by a number of different, probably vicarious, species: *C. laticurvatum, C. miocaenicum, C. bavaricus,* and others. Among the suoids, some paleochoerids remained from the late Oligocene (*Palaeochoerus*). However, the dominant genera of this family in the early Miocene were the more-advanced hyotherines, which established a presence throughout Eurasia. They included such old forms as *Xenohyus* and new genera as *Hyotherium* and *Aureliachoerus.* These hyotherine suids were typical western and central European medium-size pigs (about 60 kg) retaining a generalized morphotype, with bunodont-brachydont teeth indicating a broad, omnivorous diet. A peculiar example is *Lorancahyus,* a

FIGURE 4.2 *Reconstructed life appearance of the early Miocene rhinoceroses* Protaceratherium minutum *and* Diaceratherium aurelianense

These two rhino species, whose remains occur together in fossil sites such as Artenay, France, showed highly contrasting sizes and body proportions. While the physique of *Protaceratherium minutum*, in the foreground, corresponded to the typical, primitive "running" rhinoceroses, *Diaceratherium aurelianense* was already a large and specialized member of the Teleoceratinae, a group of robust and short-limbed rhinos that may have had partially amphibious habits. While other teleoceratines were hornless, *Diaceratherium* had a small, split knob on top of its nasals, indicating the presence of a tiny, perhaps double horn. Reconstructed shoulder height of *Diaceratherium:* 1.24 m.

FIGURE 4.3 *Reconstructed life appearance of the cainotherid* Cainotherium laticurvatum
An especially good sample of fossils of this species comes from Saint-Gérand-le-Puy, where almost every part of the skeleton is represented. This small artiodactyl had long hind limbs relative to the forelimbs, a feature that, combined with its tiny body size, has led some authors to claim that it was a sort of ecological parallel of modern rabbits and hares. Reconstructed shoulder height: 15 cm.

small peccary-like suid that developed high-crowned, tubular cheek-teeth similar to those of the order of the Tubulidentata (which includes today's aardvark from Africa). In fact, the first discoveries of this group of suids at the middle Miocene locality of Córcoles, Spain, were thought to be true tubulidentates. However, more abundant material coming from Loranca have clearly shown that the tubular pattern and absence of cusps and cuspules in the molars was simply a case of homoplasy with other ungulates adapted to ingest hard vegetation or even soil grains when eating, such as the tubulidentates from Africa or the edentates from South America.

Ruminants of the moschoid group dominated the earliest Miocene ungulate communities. The moschoids, first represented by the Oligocene survivors *Dremotherium* and *Bedenomeryx*, experienced a quick diversification during the earliest Miocene, splitting into a number of genera. Some of them, such as *Amphitragulus* and *Oriomeryx*, retained the baseline characteristics of these Oligocene genera. Others, however, like the species *Bedenomeryx truyolsi* from Cetina de Aragón, attained larger sizes (this species was, in fact, the last representative of the genus). In the opposite direction, *Pomelomeryx* diminished in size. All these genera bore brachydont dentition adapted to browse soft vegetation, including fruits and leaves, and even carrion. However, one of the last representatives of the genus *Dremotherium, D. cetinensis* from Cetina de Aragón, developed a specialized dentition, comparable to that of the first cervids, displaying more hypsodont cheek-teeth than any other ruminant of its time, probably as an adaptation to ingest harder vegetation.

This general pattern of persistence of late Oligocene lineages into the earliest Miocene times also occurred among the rodents. The last theridomorphs of the genus *Issiodoromys* were still found in some oldest Miocene localities of western Europe, and at this time this line of theridomorphs attained its highest levels of hypsodonty, with molar crowns about 8 mm high. The four other most abundant families of early Miocene rodents—cricetids, glirids, eomyids, and sciurids—also had remnant Oligocene representatives. The earliest Miocene squirrels, for instance, exhibited a remarkable stability, with the persistence of the conservative Oligocene genus *Palaeosciurus*. In contrast with the Oligocene species *P. goti*, the early Miocene *Palaeosciurus* species (*P. feignouxi*) were arboreal forms, as illustrated by their more elongated radius and tibia (figure 4.1).

The Oligocene–Miocene transition brought a significant decline among the cricetids, though, with the extinction of a number of Oligocene genera such as *Pseudocricetodon* and *Adelomyarion*. Only the members of the genera *Eucricetodon* and the bizarre *Melissiodon* persisted among the early Miocene faunas.

Among the glirids, the Eocene lineage of *Gliravus* ended, but most late Oligocene genera, such as *Peridyromys, Bransatoglis,* and *Microdyromys,* crossed the Oligocene–Miocene boundary. In fact, this group flourished and diversified in the early Miocene into a number of genera and species, most of them persisting until the beginning of the late Miocene. The early Miocene strata record a significant increase in the diversity of this family, with new genera such as *Miodyromys, Pseudodryomys, Altomiramys,* and *Glirudinus.* This diversification not only affected the taxonomic level but also involved the diversity of dental patterns present until that time in the family. From the rather simplified morphologies derived from *Gliravus,* throughout the Miocene a number of lineages simplified or, more frequently, increased the complexity of their dental patterns and increased the number of ridges of the molar crown.

A different picture emerges for the two other families persisting from the Oligocene: the eomyids and the sciurids. The small eomyids of the genus *Rhodanomys* persisted in the early Miocene and prolonged the evolutionary trends developed by the late Oligocene representatives of the genus by increasing their hypsodonty and simplifying their dental pattern. Before their extinction during the early Miocene, the last members of this lineage, commonly placed in a separate genus, *Ritteneria,* displayed cheek-teeth composed of a single pair of transverse ridges.

In the context of the subtropical conditions of these early Miocene faunas, the bat representation again attained high diversity levels in Europe, the tropical families of the molossid (or mastiff bats) and hipposiderids (or leaf-nosed bats) becoming common at some sites like Montaigu and Bouzigues, France. As at Messel during the middle Eocene, some bats like the molossid *Tadarida stehlini* developed fast flying in the open spaces of the high arboreal strata, while others like the hipposiderid *Hipposideros bouziguensis* practiced a reduced wing-loading flight close to the ground or the water surface (figures 4.4–4.7).

Within the predator guild, nimravids and amphicyonids represented the medium-size to relatively large carnivores. After a gap during the late Oligocene, the catlike nimravids reappeared during the early European Miocene, represented by the genus *Prosansanosmilus* (*P. peregrinus*). Like the Oligocene dirk-toothed nimravids, *Prosansanosmilus* displayed laterally compressed and serrated upper canines that were much longer than the lower ones. During the late to early Miocene, this genus achieved a wide geographic range, from Spain (Bunyol) to France (Artenay, Baigneaux, Bézian) and Germany (Langenau, Württemberg). In fact, the nimravids maintained a remarkable stability throughout the Miocene, probably in relation to a low speciation rate.

The amphicyonids developed quite differently, during the early and middle Miocene diversifying into a wide array of forms, ranging from omnivorous generalists to hypercarnivorous species. Some of the small unspecialized amphicyonids of late Oligocene origin, such as *Haplocyonoides* and *Cynelos,* persisted into the early Miocene. *Cynelos,* a late Oligocene–early Miocene holdover, was rather common in Europe, where it is found in several early Miocene sites from western Europe (Artenay, Bézian, Wintershof-West, Baigneaux, Buñol, and Erkertshofen). At least some species were rather cursorial, with gracile distal limbs and some degree of digitigradism, away from the basic robust body plan of most of the amphicyonids. This was the case, for instance, with *Cynelos schlosseri,* of about 23 kg. A larger species was *C. helbingi,* which weighed between 60 kg (female) and 90 kg (male) and has been found in several western European localities, from the Vallès-Penedès Basin and Bunyol, Spain, to Wintershof-West and Erkertshofen, Germany.

While the Oligocene amphicyonids resembled canids in their dentition, during the Miocene another group developed more ursidlike characteristics, such as large sizes and robust canines (with double cutting edges in the upper canines). The most striking species was *Amphicyon giganteus,* recorded in a number of localities from the Iberian Peninsula to Austria (figure 4.8). The males of this species could attain 317 kg, while the females were much less robust, with an esti-

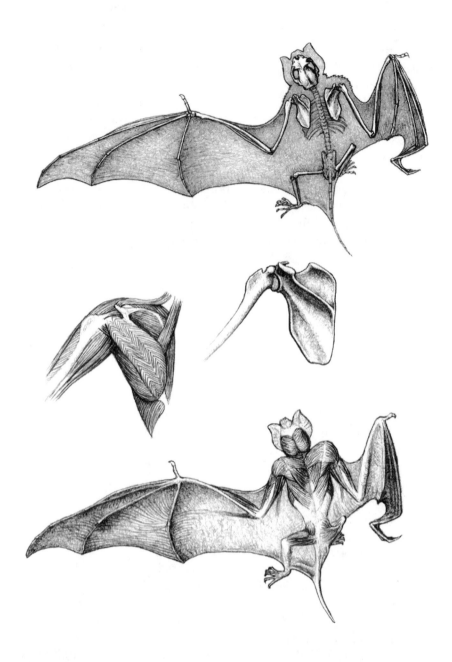

FIGURE 4.4 *Skeleton and reconstructed musculature of the molossid bat* Tadarida stehlini

Many features in the skeleton of this early Miocene bat show that it was as well adapted for high-altitude and high-speed flight as modern members of the Molossid family, which have been called "the swallows of the bat world." The double articulation between the humerus and the scapula allowed a more efficient vertical rotation at the shoulder joint, while the insertions of such shoulder muscles as the biceps brachii were modified to allow the best leverage for powerful wing beating. Such features are known thanks to the excellent preservation of many *Tadarida* fossils from the early Miocene site of Saint-Gérand-le-Puy.

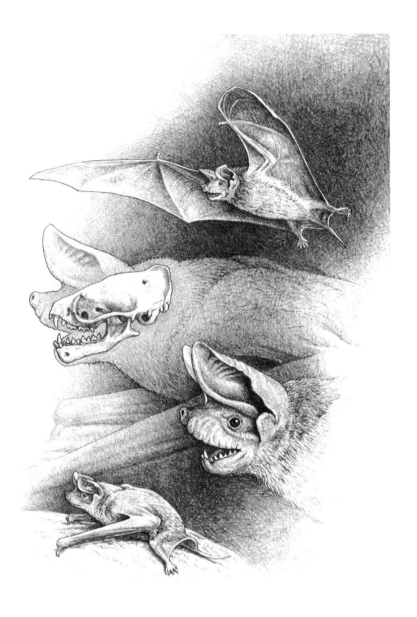

FIGURE 4.5 *Skull and reconstructed life appearance of the bat* Tadarida stehlini *in flight and on the ground*

Modern members of the bat family Molossidae display unadorned muzzles, and the skull morphology of *Tadarida stehlini* shows that the same was true for that species, since the nasal area lacks the concavity associated with leaflike nose appendages found in other bat families. There is every likelihood that, like modern molossids, *T. stehlini* emitted its ultrasounds through its open mouth during flight. Since the skeleton shows the same adaptations to high-speed flight, it is likely that the external ears had the same highly aerodynamic design as those of modern molossids. The structure of the robust hind limbs shows that, again like modern molossids, *T. stehlini* would have been a proficient (for a bat) quadrupedal walker and would rarely, if ever, have hung free while resting.

FIGURE 4.6 *Comparison between the wing proportions of* Tadarida stehlini *and* Hipposideros bouziguensis

These two bat species, found in fossil sites of similar age from the early Miocene in France, show different wing proportions—implying different flight styles and feeding ecologies. The molossid *Tadarida,* shown at top, with its long, narrow wings, would have flown high above the forest trees, catching fast-flying insects with its mouth open wide in swallow fashion. *Hipposideros,* in contrast, had shorter, wider wings well suited for maneuvering among the branches and catching insects at lower heights within the forests.

mated body mass of about 157 kg. They developed robust canines with two strong vertical enamel ridges and large third upper molars, which distinguished them from other species of the genus. The heavy carnassials with horizontal abrasion were probably used for bone-crushing, an interpretation supported by the high sagittal crests in the skull, which housed a powerful musculature likely used for breaking bones. All these adaptations suggest that the amphicyons were probably occasional scavengers, at a time when true scavenging specialists were still absent. Despite these scavenging adaptations, the body plan of *Amphicyon* indicates active hunting, and they were probably agile predators practicing the solitary stalk-pounce hunting style of today's felids. Their long tails were probably used for balance during the pounce, as with modern

FIGURE 4.7 *Skull and reconstructed life appearance of the hipposiderid bat* Hipposideros bouziguensis, *in flight and resting*

Well-preserved fossils of this bat species from the site of Bouzigues, France, clearly show the similarities in structure with modern leaf-nosed bats. Like them, *Hipposideros bouziguensis* had a hollow nasal area in the skull, where the leaflike nose appendages would have fitted. Such structures are thought to focus and concentrate the sound waves that are emitted from the nostrils. *H. bouziguensis* had relatively feeble hind limbs, well adapted for hanging free during rest but rendering the bat a poor quadrupedal walker.

FIGURE 4.8 *Reconstructed life appearance of the creodont* Hyainailouros sulzeri, *drawn to scale between the giant amphicyonid* Amphicyon giganteus *and the extant gray wolf,* Canis lupus

Hyainailouros was one of the largest members of the Creodonta, and the skull of the closely related (if not congeneric) *Megistotherium* from northern Africa measures over 60 cm in length, larger than that of any other creodont or any modern carnivore. Although complete skulls of *Hyainailouros* are not known, the size of the dentition is at a par with that of *Megistotherium,* but one must take into account that these giant creodonts, like all members of the family Hyaenodontidae, had very big skulls for their body size. The skeleton of *H. sulzeri* is well known thanks to a fine specimen from the site of Chevilly, France, which shows an animal with relatively short, robust legs and long body and tail. Compared with the contemporary amphicyonid *Amphicyon giganteus,* the giant creodont had a considerably larger head, but a smaller body. The modern wolf is the biggest extant carnivore whose dentition shows a combination of meat-slicing and bone-cracking adaptations comparable to that of *Hyainailouros* and *Amphicyon,* but it would, of course, be dwarfed by both Miocene meat predators.

felids. Another large early Miocene species was *Euroamphicyon olisipo-nensis,* widely present in several localities from western Europe (for instance, Quinta do Narigao, La Retama, and Buñol). Although smaller (about 150 kg), this amphicyonid was similar in many respects to *A. giganteus,* with reduced premolars and well-developed molars.

A third group of early Miocene amphicyonids included fully carnivorous forms. Among them, the Oligocene genus *Ysengrinia* persisted into the early Miocene in Spain. But the range of hypercarnivorous amphicyonids broadened considerably during the early Miocene with new forms such as *Pseudocyon sansaniensis,* a large species (120 to 130 kg) that displayed narrow molars badly adapted to bone-crushing. Another species of this genus, *P. caucasicus,* was a hypercarnivorous amphicyonid that had lost the anterior premolars and developed narrow, slicing car-nassials. At the end of the early Miocene, *Agnotherium grivense*, a large amphicyonid of about 170 kg, became fully hypercarnivorous; this spe-cies developed carnassials specially adapted to slice meat, as indicated by their vertical tooth wear. The members of *Pseudocyon* displayed a body plan different from that of other amphicyonids, with more cursorial and less graviportal limbs, probably adapted to pursuit hunting.

In contrast, a group of small early Miocene amphicyonids, such as *Ictiocyon socialis* and *Pseudarctos bavaricus*, developed omnivorous tenden-cies. These species have been found at several early and middle Miocene sites in western and central Europe. They were small amphicyonids of less than 20 kg (about 9 kg in the case of *P. bavaricus*). Their postcar-nassial molars, however, presented large occlusal surfaces that provided a large grinding area. Viranta (1996) has suggested a diet similar to that of today's lesser panda (*Ailurus fulgens*), which is composed of bamboo sprouts, grasses, roots, fruits, nuts, and, occasionally, small vertebrates, eggs, and insects. The postcranial skeleton indicates good arboreal abil-ities (Ginsburg 1999).

THE PROBOSCIDEAN EVENT AND THE GOMPHOTHERE LAND BRIDGE

Despite the climatic stability of the early Miocene, an important tectonic event disrupted the evolution of the Eurasian faunas during this epoch. About 19 million years ago, the graben system along the Red Sea Fault, active in the south since the late Oligocene, opened further, as evi-denced by the presence of shallow-water sediments along the coast of Sudan and in the eastern desert to the north. Consequently, the Arabian plate rotated counterclockwise and collided with the Anatolian plate. The marine gateway from the Mediterranean toward the Indo-Pacific closed, and a continental migration bridge (known as the Gomphothere Bridge) between Eurasia and Africa came into existence.

This event had enormous consequences for the further evolution of the terrestrial faunas of Eurasia and Africa. Since the late Eocene, Africa had evolved in isolation, developing its own autochthonous fauna. Part of this fauna consisted of a number of endemic Oligocene survivors, such as anthracotheres, hyaenodonts, and primates, for which Africa had acted as a refuge. However, as in other southern continents, such as Australia and South America, the isolation of the African plate fa-vored the evolutionary radiation of several lineages of autochthonous ungulates, which are often assembled in the clade Tethytheria. The teth-ytherians include a number of ungulate groups of African origin, such as the hyraxes, aardvarks, embrithopods, demostylians, sirenians, and proboscideans. The first evidence of an African–Eurasian exchange was the presence of the anthracothere *Brachyodus* in a number of early Miocene sites in Europe. This entry of the anthracotheres into Europe was a rather short-lived event, for they did not reach the middle Miocene, although they survived in southwestern Asia and Africa. On the contrary, a second dispersal event from Africa, that of the gomphothere and deinothere proboscideans, had much more lasting effects.

Today we can easily identify any proboscidean by its long proboscis and tusks. However, the primitive proboscideans from the African Eocene had a completely different appearance and are hardly recognizable as the ancestors of today's elephants. Instead, they were hippolike semiamphibious ungulates with massive, elongated bodies supported by rather short legs. One of the first was *Moeritherium,* from the late Eocene of El Fayoum, Egypt. *Moeritherium* had bunodont and brachydont cheek-teeth, adapted to ingest soft vegetables and fruits. However, this form already showed a tendency toward elongation of the upper and lower incisors, which would become the tusks of the late Oligocene and Miocene proboscideans. By the middle Oligocene, the genera *Palaeomastodon* and *Phiomia* from El Fayoum exhibited the features that characterized the mastodons during the following millions of years: bunodont teeth formed of a number of rounded hills, retracted nasals (an indication of the presence of a trunk), and strongly developed upper and lower incisors forming two pairs of tusks.

The first proboscideans entering Europe were the so-called gomphotheres, represented by the genus *Gomphotherium,* which dispersed worldwide during the early Miocene from Africa to Europe, Asia, and North America (figures 4.9 and 4.11). *Gomphotherium* was the size of an Indian elephant, about 2.5 m high at the withers. Its skull and dentition, however, were different from those of modern elephants. *Gomphotherium*'s skull was long (it is, therefore, known as a "longirostrine" mastodon) and displayed not two but four tusks, one pair in the upper jaw and the other pair at the end of the lower jaw. The upper pair was directed downward, instead of upward, as in the more recent elephants. The second pair of tusks, in the mandible, was straight and smaller. In most of the gomphotheres, the presence of elongated mandibles with a long symphysal region compensated for the difference in size between the upper and the lower tusks. This tendency reached its extreme in *Amebelodon* and *Platybelodon,* two Asian genera in which the lower tusks were flat and transformed into a sort of shovel, probably used to dig into the mud. The dentition of the gomphotheres was bunodont and low-crowned, each tooth bearing a number of paired hills that resembled breasts (hence the name *Mastodon,* or "breast-tooth," coined by the great French paleontologist Georges Cuvier in 1806). Characteristically, the third molars of *Gomphotherium* bore three pairs of hills—thus the

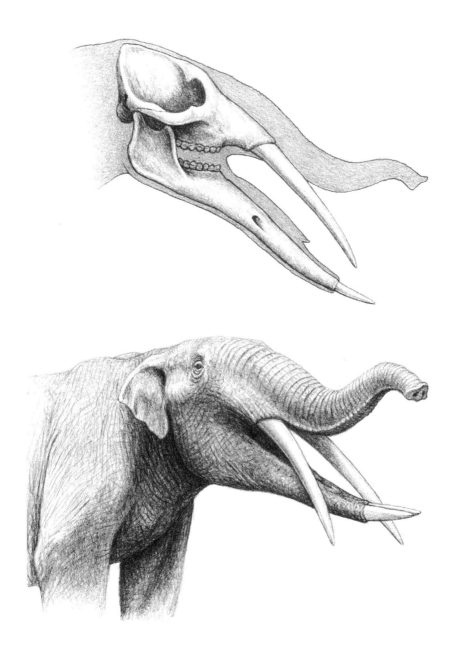

FIGURE 4.9 *Skull and reconstructed head of the mastodon* Gomphotherium angustidens

This reconstruction is based on a spectacular skull found at the site of Tetuán de las Victorias, Spain, and shows an extreme example of the elongated mandible that characterized this species. Later species in the genus, such as *Gomphotherium stein-heimensis,* showed a trend toward shortening the mandible, foreshadowing the short-jawed morphology of the middle Miocene genus *Tetralophodon.* The function of such a long mandible is not readily apparent, although it was most likely used in concert with the trunk for gathering food.

generic assignment *Trilophodon* (teeth with three ridges), which was also used for *Gomphotherium*.

Accompanying *Gomphotherium* to Europe during the early Miocene was a second group of mastodons, those of the genus *Zygolophodon*. Basically, members of *Zygolophodon* differed from the gomphotheres in their peculiar cheek-teeth, which are often known as "zygodont" dentition. In the zygodont molars, the typical "mastodon" hills were transformed into sharp ridges, with more clearly defined cutting edges. This difference from the gomphotheres probably reflected a diet composed of tougher vegetation.

Shortly after the entry of *Gomphotherium* and *Zygolophodon,* a third proboscidean group, the deinotheres, successfully settled in Eurasia. Unlike the previous genera, the deinotheres were not elephantoids but represented a different, now totally extinct kind of proboscidean. Members of this group had only two strong tusks, not placed in the upper jaw but at the end of the mandible. Moreover, this lower pair of tusks was recurved downward. The molars were simple, formed of two cutting ridges; as in the zygodont mastodons, this dentition is interpreted as an adaptation to browsing leaves and tough vegetation. In contrast to that of the gomphotheres, the origin of the deinotheres is not yet clear. They were probably related to a group of primitive African proboscideans known as the barytheres. The barytheres were rather small hippolike proboscideans bearing a bilophodont dental pattern (as did the deinotheres) and two short tusks at the end of the mandible. Like *Moeritherium,* barytheres have been found in the late Eocene sediments of El Fayoum. The first deinotheres in Europe, assigned to the genus *Prodeinotherium,* were relatively small proboscideans about 2 m high at the withers (figure 4.10). However, the middle and late Miocene representatives of this group, included in the genus *Deinotherium,* were tall forms reach-

....

FIGURE 4.10 *Skull with anterior skeleton and reconstructed musculature of the head and neck of the proboscidean* Prodeinotherium bavaricum

The shape of muscle insertion areas in the neck vertebrae and posterior skull of deinotheres reflects a marked specialization for enhanced movement in the vertical plane. Compared with those of a modern elephant, the muscles that pull the head up and those that bring it down can act through a wider arc. This was very likely related to the action of the downward-pointing tusks, although the precise function of the tusks remains a matter of debate. Since the deinotheres were clearly obligate browsers—as indicated by the morphology of their cheek-teeth—it seems likely that the tusks were used in concert with the trunk to gather foliage from the branches of trees. The shape of the nasal area clearly indicates that there was, indeed, a well-developed proboscis, although some primitive features in that area suggest that the muscular control of trunk movement was not as sophisticated as it is in modern elephants, and it has been suggested that the trunk itself would have been relatively shorter.

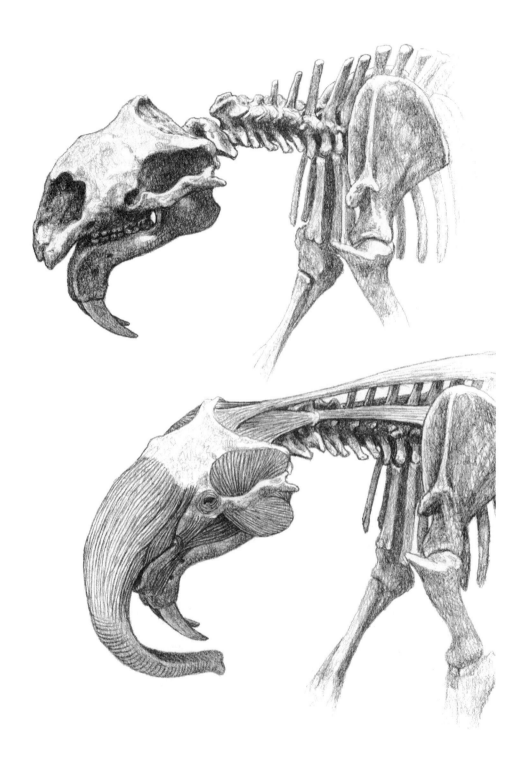

ing 4 m in height at the withers (the largest terrestrial mammals of their time) (figure 4.11).

THE EASTERN IMMIGRANTS

The dispersal of not only the African proboscideans but also many eastern immigrants contributed to a significant increase in the diversity of the impoverished early Miocene terrestrial biotas. The entry of this set of immigrants probably led to the extinction of a number of late Oligocene and early Miocene survivors, such as tapirids, anthracotherids, and primitive suids and moschoids. In addition to the events that affected the Middle East area, sea-level fluctuations enabled short-lived mammal exchanges across the Bering Strait between Eurasia and North America, permitting the arrival of the browsing horse *Anchitherium* in Eurasia and the chalicotheres and various mustelids and rodents in North America.

Anchitherium was a member of the anchitherines, a group of North American horses that experienced a significant evolutionary radiation during the Oligocene (figure 4.12). They were larger than the primitive Eocene equids and had developed more-derived dentition, with molarlike premolars and an increased tooth-crown area (MacFadden 1992). However, they still had low-crowned (brachydont), lophodont teeth adapted to browsing leaves. Their limbs were also adapted to locomotion on soft substrates, still retaining two lateral, functional toes. Despite the diversity of this group in North America (*Mesohippus*, *Miohippus*, and others), only *Anchitherium* entered Europe. *Anchitherium* was a medium-size anchitherine, of about 1 m or less at the withers, that rapidly dispersed over the whole of Eurasia, from China to Spain. Widely used for biostragraphic purposes, the dispersal of *Anchitherium* was the first of a number of similar isolated events undergone by North American equids that entered Eurasia and rapidly spread on this continental area.

Another group of perissodactyls that entered Europe at this time were the chalicotheres, represented by the genera *Metaschizotherium* and *Phyllotillon* (plate 5). Both belonged to the subfamily Schizotherinae, characterized by their semihypsodont, specialized dentition. Unlike those of the more common middle and late Miocene chalicotheres, their forelimbs were still capable of conventional locomotion over the ground. *Metaschizotherium* and *Phyllotillon* were probably part of a wide dispersal event that resulted in a considerable widening of the geographic range of this group, from Asia to Europe and Africa, on the one hand, and from Asia to North America, on the other (attested by the first appearance of *Moropus* in North America). Nevertheless, the

FIGURE 4.11 *Comparison between* Gomphotherium angustidens *and* Deinotherium giganteum

The size and proportions of *Deinotherium giganteum* as shown in this illustration are based on a set of complete limb bones from the site of Cerecinos del Campo, Spain. They indicate an approximate shoulder height of 3 m, but some specimens from other sites suggest taller animals reaching 4 m. The reconstruction of *Gomphotherium angustidens* is a combination of the skull and mandible from Tetuán de las Victorias, with the postcranial remains from the sites of Artenay and Sansan, France, which indicate a shoulder height of about 2 m.

presence of *Metaschizotherium* during the early Miocene in Europe may also be a case of local evolution from the Oligocene *Schizotherium*.

A third group of perissodactyls, the rhinoceroses, also experienced a significant turnover during the early Miocene. The lineages of the small rhinoceroses that survived the Oligocene–Miocene boundary (*Protaceratherium*, *Menoceras*) came to an end. The primitive aceratherines of the genus *Mesaceratherium* gave way to the more-derived species of the genus *Plesiaceratherium* (*P. fahlbuschi, P. mirallesi*). They were slender aceratherines already displaying the characteristics to be found in the later middle and late Miocene members of the tribe (*Hoploaceratherium, Aceratherium*): long, hornless nasals and long, flattened, outcurving lower incisors (which were less developed in the females). The upper incisors had lost their shearing function and finally disappeared. In

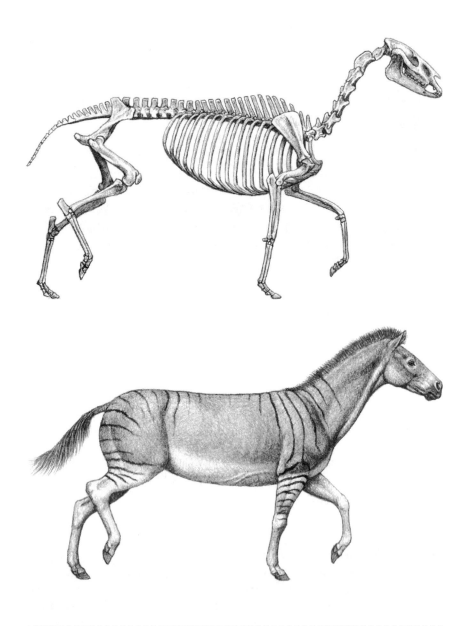

FIGURE 4.12 *Skeleton and reconstructed life appearance of the equid* Anchitherium aurelianense

The abundant remains of this horse found at the famous site of Sansan show a curious mixture of primitive and advanced features. The animal was still functionally tridactyl, but the third digit clearly supported most of the weight. The hind limbs were relatively long and the forelimbs short, resembling the proportions of earlier equids and reflecting an adaptation to locomotion in forested habitats. Such limb proportions, together with a relatively long back and neck and a small head, contributed to give the animal a vaguely deerlike look. The head, in turn, had a shorter snout and a longer brain case than those of modern horses, with the eyes placed nearer to the center of the skull instead of far back, as in today's species.

their place, a prehensile lip for snapping off leaves and twigs was present, as suggested by the retracted nasals.

But the most significant event in the early Miocene rhinocerotid world was the widespread presence of the teleoceratines. This group attained a certain diversity in Europe, with two genera, *Diaceratherium* and *Prosantorhinus,* derived from the late Oligocene *Brachydiceratherium.* *Diaceratherium brachypus* was a large, short-legged teleoceratine of about 1,500 kg with relatively high-crowned molars (figure 4.2). Its skull was short, with a small, split knob on top of its nasals, indicating the presence of a tiny, perhaps double horn. In contrast, *Prosantorhinus germanicus* was a small teleoceratine with shortened legs and more brachydont molars than *D. brachypus* but displaying a more-developed nasal horn. Both probably lived close to swampy, limnetic environments, the associated soft aquatic vegetation becoming their main resource (although *Prosantorhinus* probably included a certain quantity of reed grasses in its diet; Heissig 1999b).

Beyond the perissodactyl world, the addition of several lineages of artiodactyls that persisted during most of the Miocene considerably enriched the browser guild. One of the most characteristic lineages was the listriodontine pigs, which appeared during the early Miocene in Europe with the genus *Bunolistriodon.* Like other listriodontines, *Bunolistriodon* was a medium-size to large suid of about 90 kg showing an incipient trend toward developing semilophodont molars (plate 5). This probably indicates a deviation from the omnivorous diet common among the early Miocene suids toward a fully folivorous diet. Their most characteristic feature was the peculiar morphology of their incisors, which were wide and short and probably adapted to serve as a cropping device.

The paleochoerids persisted at this time in the form of small pigs of the genus *Taucanamo,* first known from the French locality of Artenay. *Taucanamo* was a small peccary-like pig (about 12 kg) with large and elongated premolars showing a tendency to develop lophodont teeth. This dental morphology, with high cusps and variable lophodonty, resembles that of some cercopithecoid primates and tragulids and probably indicates that *Taucanamo* browsed in a forest biome.

Among the archaic artiodactyls, another significant event was the appearance in Europe of the first tragulids of the genus *Dorcatherium* (plate 7). The tragulids, today represented by the mouse deer *Tragulus meminna* and *Hyaemoschus aquaticus,* are small ruminants that live in the dense tropical forests of Africa and eastern Asia, close to rivers and watercourses that provide refuge in case of sudden attack by predators. Primitive ruminants, they retain a number of features absent in the more-derived forms, such as the cervids and the bovids. Like the moschoid ruminants, they lack any kind of cranial appendages, having instead a pair of long canines that are longer in the males than in the females (a characteristic also present in several moschoids). Moreover, they retain a primitive limb structure, with four well-developed toes on each foot (the two central metapodials are not yet fused in the cannon bone). *Dorcatherium* was nearly identical in all respects to *Hyaemoschus* and probably developed a similar lifestyle. The presence of *Dorcatherium* in the earliest Miocene levels of Songhor in eastern Africa indicates that the Gomphothere Bridge not only allowed movement of animals from Africa to Europe but also enabled a number of Eurasian taxa to settle in Africa.

The early Miocene faunal turnover deeply affected not only the large and medium-size herbivores, but also the very small ones. Like several ungulates, a number of rodent survivors from the late Oligocene declined and disappeared during this period. First was the extinction of the last theridomorph, *Issiodoromys*, followed by the eomyids of the *Rhodanomys-Ritteneria* group, which came to an end after millions of years of successful evolution. This was not the end of the eomyid rodents in Europe, however, since a second lineage, that of *Pseudotheridomys* and *Ligerimys*, continued in the early Miocene. *Pseudotheridomys* had relatively complicated molars formed of a number of parallel ridges, while its descendant, *Ligerimys*, evolved a simpler dental pattern based on four main crests. *Ligerimys* was common among the early Miocene rodent faunas and appears to have been associated with limnetic, wet environments. In addition, the ochotonid lagomorphs, rare during the Oligocene, became common among the small mammalian faunas in the early Miocene. Particularly, two genera in this family, *Piezodus* and *Lagopsis*, are usually present in terrestrial communities of this time (*Piezodus* gave rise in the early to middle Miocene to *Prolagus*, the most common and almost unique lagomorph during most of the Miocene).

But the most profound renewal among rodents was that of the cricetids, or hamsters, with the gradual decline and final disappearance of those of the genera *Eucricetodon* and *Melissiodon*. During a short interval, between the dispersal events of the proboscideans *Gomphotherium* and *Prodeinotherium*, the cricetids were almost absent from the European paleofaunas (only scarcely represented by the bizarre *Melissiodon*, the so-called Cricetum vacuum). But almost coinciding with the entry of the deinotheres from Africa, several waves of new cricetid immigrants of eastern origin entered Europe: *Democricetodon, Eumyarion, Megacricetodon, Cricetodon, Neocometes*, and others. Some of these genera lie at the origin of lineages that would persist over millions of years, such as *Cricetodon* (the first member of a lineage that ended in the late Pliocene) and *Democricetodon* (probably the origin of today's hamsters). Others, such as *Megacricetodon*, although shorter-lived, were opportunistic animals showing an extraordinary speciation rate and were widely present among all early to middle Miocene biota. The origin of most of these genera can be traced to western Asia, and some of their ancestors, including those of the archaic genus *Eumyarion*, have been identified in the Levant (Anatolia) and eastward. The relatively sudden entry of all these hamsters during the early Miocene led to unprecedented levels of rodent diversity. This increase in diversity was even greater because the other dominant rodent family, the glirids, or dormice, did not decline under the impact of these new immigrants but, on the contrary, increased their diversity with the addition of new genera.

THE REVOLUTION OF THE HORNED RUMINANTS

The main faunal revolution in the herbivore community during the early Miocene, despite the increased diversity of the cricetids and glirids, was the appearance and spread of the first horned ruminants. The presence of cranial appendages is easily recognized in a variety of ruminant families included in the Pecoran clade, such as modern cervids, giraffids, bovids, and antilocaprids, and some extinct families, such as the Eurasian palaeomerycids and the American dromomerycids. The diversity of these cranial appendages (ossicones, antlers, and horns) and their presence in several ruminant families suggest independent origins in separated areas rather than a unique source.

In Europe, the first ruminants bearing cranial appendages appeared during the early Miocene in Spain (Ramblian), represented by primitive giraffids of the genera *Teruelia* (from the locality of Navarrete del Río) and *Lorancameryx* (from the locality of Loranca). The giraffids have two or more appendages right above the orbits but lack the typical horn sheaths present in the bovids. These primitive giraffoids did not display the long necks of today's giraffes (in fact, this is a unique characteristic of *Giraffa*, absent in most of the remaining members of the family, including the okapi). *Teruelia* and *Lorancameryx* had well-developed ossicones (present only in the males) and shared the basic features that characterized the primitive giraffids. *Lorancameryx* was closely related to *Teruelia* but had heavy, pachyostotic limb bones.

The presence of *Teruelia* and *Lorancameryx* in the early Spanish Miocene was certainly the result of a dispersal event from the east, and both genera were probably related to the primitive giraffids of the subfamily Progiraffinae, which include *Progiraffa*, from the Bugti Beds in the Siwaliks Hills, Pakistan, the oldest giraffid known. Like *Progiraffa*, *Teruelia* and *Lorancameryx* retained some primitive features, such as upper canines. But they lacked other characteristics present in today's giraffids, such as bilobate lower canines.

The geographic distribution of the first giraffids suggests that this group originated in Asia to the south of the Alpine belt and dispersed during the early Miocene to western Europe and northern Africa, where they were represented in the early to middle Miocene by a variety of genera: *Canthumeryx, Climacoceras, Prolibytherium,* and *Palaeotragus.* The origin of the first giraffoids is still unknown, but a plausible candidate would be *Andegameryx,* a primitive pecoran genus of uncertain affinities that was widely present during the early Miocene in western Europe (Cetina de Aragón, Valquemado, Beilleaux, Faluns de Touraine, Anjou) and the Indian subcontinent. Although *Andegameryx* lacked the characteristic appendages of bovoids and giraffoids, this genus had some of the derived dental features that are seen in these groups (Morales et al. 1986).

The second evolutionary radiation of horned ruminants was that of the deer, or cervids. The cervids differ from other horned ruminants in their peculiar cranial appendages, antlers, which usually only the males grow. In today's cervids, the base of the antler, the pedicle, is a permanent structure attached to the skull, while the antler is deciduous, growing yearly. Close to the base of the antler, a burrlike structure, the rose, defines the boundary between the permanent and the deciduous part of the antler. The first cervids entered Europe during the early Miocene, at the same time as the first proboscideans. These primitive cervids, usually included in the family Lagomerycidae, were small animals that retained some primitive characteristics present among their close moschoid ancestors, such as a pair of long upper canines in the males. Their cranial appendages were formed of a long and straight pedicle (more than two-thirds of the total length) without a burr, ending in a much reduced antler (for which the term *protoantler* has been coined). The shape of these antlers varied from one genus to other, and the most primitive forms, such as *Ligeromeryx* and *Procervulus,* bore a pair of straight, slightly divergent pedicles about 15 cm long that ended in a two-pronged fork. They lacked any kind of burr, but some form of occasional casting probably existed (plate 5). *Lagomeryx* displayed similar straight pedicles, but they ended in a multipointed antler instead of the two-pronged fork of other genera. The antlers of the small cervid *Acteocemas* did not look much different from those of *Procervulus,* with a long, straight pedicle and a two-pronged fork at its end, but *Acteocemas* had a small burr at the base of the antler, indicative, perhaps for the first time, of a true, probably cyclic, process of casting. Therefore, *Acteocemas* seems the plausible ancestor of the middle Miocene cervids such as *Dicrocerus, Euprox,* and other later Miocene genera.

The third kind of early to middle Miocene horned ruminants was the group of paleomerycids, heavy, long-legged, massive animals that could attain a weight of 350 to 500 kg. *Palaeomeryx,* one of first known

PLATE 1 *A forest at the site of La Boixedat, early Eocene*
Left to right: the adapid primate *Agerinia;* the
ophiodontid perissodactyl *Lophiodon;* the condylarth *Phenacodus;* and the creodont *Proviverra.*

PLATE 2 *A lakeshore at the site of Roc de Santa, late Eocene*

Left to right: the selenodont artiodactyl *Xiphodon;* the paleotherid perissodactyl *Paleotherium;* and the creodont *Hyaenodon* feasting on the carcass
of a paleotherid.

PLATE 3 *A swamp at the site of Tárrega, Oligocene*
Left to right: the crocodile *Hispanochampsa;* the cainotherid *Cainotherium;* the anthracotherid *Elomeryx;* the musteloid carnivore *Plesictis;* and the rodent *Theridomys.*

PLATE 4 *A water hole in a semiarid environment at the site of Carrascosa*

Left to right: the crocodile *Hispanochampsa;* the nimravid carnivore *Nimravus;* the cainotherid *Cainotherium;* the "running rhinoceros" *Egyssodon;* and the piglike artiodactyl *Methriotherium.*

PLATE 5 *A riverine woodland at the site of Buñol, early Miocene*

Left to right: the pig *Bunolistriodon;* the hornless rhinoceros *Mesaceratherium;* the equid *Anchitherium;* the mastodon *Gomphotherium;* the early deer *Procervulus;* the horned rhinoceros *Dicerorhinus;* the chalicotherid *Phyllotillon;* and, in the foreground, the hemicyonid carnivore *Hemicyon.*

PLATE 6 *A wooded lakeshore at the site of Els Cassots, early Miocene*

Left to right: the pangolin *Teutomanis;* the cervid *Procervulus;* the felid *Pseudaelurus;* the hornless amphibious rhinoceros *Diaceratherium;* the mastodon *Gomphotherium;* the amphicyonid carnivore *Amphicyon,* dragging the carcass of a pig; the paleomericyd *Ampelomeryx;* and the small, early bovid *Eotragus.*

PLATE 7 *A gallery woodland in an arid environment at the site of Pasillo Verde, middle Miocene Left to right: the hornless rhinoceros Hispanotherium; the chevrotain Dorcatherium; the cervid Procervulus; the pig Bunolistriodon; and the paleomerycid Triceromeryx.*

PLATE 9 *A lacustrine basin at the site of Batallones, Vallesian*

Left to right: the equid *Hipparion* (*Hippotherium*); the saber-toothed felid *Machairodus;* and the mastodon *Tetralophodon.*

PLATE 10 *A riverine woodland at the site of Crevillente,*
Turolian

Left to right: the proboscidean *Deinotherium;* the equid *Hippotherium;* the
saber-toothed felid *Machairodus,* with the carcass of a *Hystrix;* the bovid
Tragoportax; the cervid *Lucentia;* and the giraffid *Birgerbohlinia.*

PLATE 11 *A water hole in open woodland at the site of Concud, Turolian*

Left to right: the saber-toothed felid *Machairodus,* with the carcass of a cervid, *Turiacemas;* the bovid *Protoryx;* a herd of small bovids of the genus *Hispanodorcas* and a group of equids of the genus *Hipparion;* the hyaenid *Lycyaena* and the small canid *Canis;* and the mastodon *Tetralophodon.*

PLATE 12 *A humid swamp near the coast at Perpignan,*
Pliocene

Left to right: the mastodon *Anancus;* the monkey *Doli-*
chopithecus; the small canid *Nyctereutes;* the aardvark
Orycteropus; the equid *Hipparion;* the tapir *Tapirus;*
the giant tortoise *Cheirogaster* (*C. perpiniana*); and the
metailurine cat *Dinofelis.*

PLATE 13 *A floodplain grassland at the site of Hué-lago, Pliocene*

Left to right: the equid *Equus stenonis;* the antilopine bovid *Gazellospira torticornis;* the bovine bovid *Leptobos elatus;* the early mammoth *Mammuthus meridionalis;* and the cervid *Croizetoceros ramosus.*

PLATE 14 *A hilly environment at the site of Atapuerca (lower TD 6 level), early Pleistocene*

A herd of hemionine horses, *Equus altidens.*

PLATE 15 *A riverine woodland at the site of Atapuerca (TD 6 level "Aurora"), early Pleistocene*

Left to right: the wild boar *Sus scrofa;* the large deer *Eucladoceros giulii;* and the early long-legged bison *Bison voigtstedtensis.*

PLATE 16 *An Atlantic gallery woodland at the site of Atapuerca (TD 8 level), middle Pleistocene*

Left to right: the hippopotamus *Hippopotamus amphibius;* the early fallow deer *Dama vallonetensis;* and the rhinoceros *Stephanorhinus etruscus.*

members of this group, was thought to be a hornless form distantly related to the giraffids. However, in 1946 the Spanish paleontologist Miquel Crusafont found in the middle Miocene of central Spain the remains of *Triceromeryx,* a *Palaeomeryx*-like form that carried two ossicones over the orbits (figure 4.13). These ossicones were straight and short, similar to those of the true giraffids. But the most striking feature of *Triceromeryx* was the presence of a third appendage at the back of the skull, prolonging the occipital bone. This posterior appendage had a Y-like shape. In the 1980s and 1990s, new discoveries of this strange group were added to the list of palaeomerycids, showing a surprising variety of occipital appendages.

At the early Miocene site of Els Casots, in the Vallès-Penedès Basin, a newly discovered genus of the group, *Ampelomeryx,* had, like *Triceromeryx,* three appendages, two over the orbits and the third in an occipital position; but the shapes of these appendages were quite different from those of *Triceromeryx* (figure 4.13; plate 6). First, the two paired ones were flat and wide and extended laterally over the orbits instead of vertically, forming a kind of eyeshade. But the most spectacular feature of *Ampelomeryx* was its unpaired posterior appendage, which was Y-shaped, as in *Triceromeryx,* but about 20 cm long, much bigger, and also flattened.

Moreover, in the middle Miocene site of La Retama, a second *Triceromeryx* species, *T. conquensis,* showed another different posterior appendage, even more spectacular than that of *Ampelomeryx.* The occipital appendage of *T. conquensis* was not Y-shaped but had the two posterior branches expanded laterally, and their extremes recurved again to the front, resulting in a sort of telephone-like structure.

FIGURE 4.13 *Reconstructed heads of paleomerycids from the European Miocene*
Left to right: Triceromeryx pachecoi, Triceromeryx conquensis, and *Ampelomeryx ginsburgi.*

The paleomerycids, therefore, appear to have been a successful group of ruminants that radiated into a variety of forms during the early and middle Miocene and attained a wide geographic range, from Spain to China. Their limb proportions were more similar to those of today's large bovids, such as buffaloes, than to those of the okapi. They probably inhabited boggy forests, standing in water like the recent elk and living on soft leaves and aquatic plants (as indicated by their brachydont teeth, similar to those of the primitive giraffids). Although once seen as giraffids, the paleomerycids are now known to have been part of an independent radiation of horned ruminants, probably derived from a moschoid cervoid like *Bedenomeryx* or *Oriomeryx* (Ginsburg 1985). Indeed, the existing evidence points to an independent, vicarious origin of giraffoids and cervoids (including the true cervids plus the paleomerycids) within the Pecora group. While the giraffoids seem to have originated in Asia, to the south of the Alpine belt (China, India, Pakistan), radiating from there to western Europe, Arabia, and Africa, the cervoids, although again immigrants to western Europe, seem to have originated in Asia to the north of the Alpine belt.

What could have been the function of the strange and almost unique occipital appendages of the paleomerycids? In the most primitive genera in the group, such as *Ampelomeryx,* the third appendage was formed from a posterior expansion of the occipital bone, close to the attachment of the powerful musculature that supports the skull in a normal position. The well-preserved specimens from Els Casots clearly show that part of this musculature was still attached to the base of the appendage, suggesting that, like the antlers of today's cervids, the appendages of *Ampelomeryx* were used in fights between males probably during breeding time. The shape of their appendages, flattened and laterally expanded in a way that could not cause any damage to the opponent, supports this hypothesis. The reduction and later evolution of these appendages in *Triceromeryx* and other middle Miocene paleomerycids suggest that they were no longer used in active fighting, but most probably had a passive display function.

The fourth part of the "early Miocene horn revolution" in Europe concerned the bovids. Like their close relatives the giraffids, the bovids bear permanent ossicones over the orbits (the horn core) that are always unbranched. Unique to this family, these ossicones are covered by a sheath made of keratin. Unlike the cervids' antlers, these horns used to be present in both males and females. The first known bovid was *Eotragus,* a small ruminant the size of the living dik-dik, or dwarf antelope, of the Neotragini tribe (plate 6). The horn cores were short and conical, placed directly over the orbits. The teeth were brachydont, indicating a diet based on soft plants, fruits, larvae, insects, and even carrion. The limb proportions were primitive, close to those of a cervid.

Eotragus is known in Europe as well as in the Kamlial beds in Turkey. However, the oldest record comes from southwestern Asia (Siwaliks, Pakistan). Like their relatives the giraffoids, the first bovids such as *Eotragus* probably originated in Asia, to the south of the Alpine belt. Later, during the early Miocene, they probably colonized western Europe and Africa simultaneously, their presence on the latter continent having been reported from Gebel Zelten (Libya) and Maboko (eastern Africa, although the age of this African locality is probably middle Miocene, as suggested by the presence of *Listriodon*).

The Carnivore Revolution

In the early Miocene, the amphicyonids were confronted with the entry of the first true ursids of the genera *Ursavus* and *Hemicyon*. *Ursavus brevirhinus,* the first member of a long-lived Miocene lineage, was a rather small bear of about 80 kg that displayed the general characteristics that we see today in this family, including a large crushing dentition adapted to a wide, omnivorous diet. A more specialized genus was *Hemicyon,* a true ursid whose appearance differed greatly from that of modern bears (figure 4.14; plate 5). Despite its large size (about 1.5 m long), it was not a heavy, robust carnivore but a slender one. Again, unlike today's bears, *Hemicyon* was not plantigrade but digitigrade, supporting all its weight on the extremes of its feet like a wolf or a lion. This suggests that *Hemicyon* must have been an active hunter and a good runner, perhaps practicing cooperative hunting. Although generally ursid in character, with a large crushing area, like that of most Miocene species its dentition showed more carnivorous specializations than that of today's bears. Even more specialized was *Plithocyon,* which showed more cursorial abilities than *Hemicyon,* with slenderer, more felidlike limb bones. Moreover, its enlarged carnassials indicate a tendency toward hypercarnivorism.

These slender ursids were not the only predators that competed with the hypercarnivorous amphicyonids and the nimravids in filling up the "large-cat" guild. As a consequence of the several faunal exchanges that took place in the early Miocene, the nimravids faced competition from true felids of the genus *Pseudaelurus,* which at that time spread over Europe together with *Anchitherium.* The earliest and most primitive species of this genus, *Pseudaelurus turnauensis* (present at several early Miocene localities, such as Wintershof-West and Loranca), was a small, catlike carnivore the size of a wild cat. It probably originated directly from the late Oligocene *Proailurus* and, like it, was arboreal, specializing in preying on the small mammals that inhabited the early Miocene subtropical forests.

In addition to the original *P. turnauensis,* this genus quickly split into two additional, usually sympatric species. The most conservative of

FIGURE 4.14 *Reconstructed life appearance of the early ursid* Hemicyon sansaniensis

The most complete remains of this species come from Sansan, but even that sample is rather fragmentary, so this reconstruction is based on a nearly complete skeleton of the closely related *Hemicyon ursinus* from the site of Santa Fe, New Mexico. The known remains of *H. sansaniensis* are virtually identical to those of the American species, showing an animal quite different in body proportions to its relatives, the modern bears. Unlike them, *H. sansaniensis* had a relatively long, flexible back, limbs with broadly catlike proportions, and digitigrade feet. *Hemicyon* would have been much fleeter of foot than any modern bear and probably more of an active predator.

these, *P. lorteti*, retained most of the characteristics of *P. turnauensis* but was slightly larger, the size of a lynx (figure 4.15; plate 6). *P. lorteti* was in many respects similar to a basic modern felid, but with shorter metapodials, indicating that it was well adapted to climb trees and lacked the cursorial abilities of its present-day relatives. The imprints left by one of these felids in the early Miocene locality of Salinas de Añana, in northern Spain, confirms this by showing that *P. lorteti* was less digitigrade than today's felids, resting on a larger foot surface when running. In this way, it resembled the forest-dwelling species of modern cats, with long backs and relatively large paws.

A third species of *Pseudaelurus*, *P. quadridentatus*, was larger than *P. turnauensis* and *P. lorteti* (about 30 kg, the size of a puma) and exhibited a trend toward longer upper canines, thus paralleling the "saber-toothed" adaptation that until that time had been characteristic of the nimravids. *P. quadridentatus* was the first true felid showing this feature and, thus, fits as a plausible ancestor for the several "saber-toothed" cats that flourished during the late Miocene and Pliocene. Until the beginning of the late Miocene, this species coexisted with different nimravid species, apparently without entering into competition with them.

Surprisingly, the oldest and most primitive species of *Pseudaelurus, P. turnauensis,* persisted until the early Turolian in Dörn-Durkheim, Germany, about 8 million years ago, surviving more than 2 million years after the last *P. lorteti* and *P. quadridentatus.* This was probably because it

FIGURE 4.15 *Reconstructed life appearance of the early felid* Pseudaelurus lorteti *and the gliding rodent* Albanensia

Skeletal remains of *Pseudaelurus lorteti* from Sansan reveal a cat similar in size and limb proportions to the modern clouded leopard of Asia—that is, an animal well adapted to climbing and occasionally catching prey in trees. Gliding squirrels like *Albanensia,* known from sites like La Grive, France, and Can Llobateres, Spain, usually would have escaped easily from such predators, thanks to their well-developed patagium, or flight membrane.

retained fully arboreal capabilities, while the other more terrestrial members of the genus had to compete with the first machairodonts. Like the amphicyonids and several other groups of Eurasian mammals, *Pseudaelurus* colonized Africa during the early Miocene across the Gomphothere Bridge, being present at the localities of Gebel Zelten, Libya, and Negev, Israel.

In the opposite direction, an African creodont, *Hyainailouros,* entered Eurasia across the Gomphothere Bridge. *Hyainailouros* is present in a number of European and western Asian localities, such as Artenay, Chevilly, and Baigneaux, France, and the Bugti Beds, Pakistan (figure 4.8). *Hyainailouros* was initially identified as an enigmatic giant carnivore on the basis of some scarce remains, but the discovery of an almost complete skeleton at Chevilly has shown that this animal was not a true carnivore but a creodont. This was a surprising conclusion, since, as we have seen in chapter 3, the creodonts declined and finally disappeared from Europe during the latest Oligocene, being replaced by the true members of the order Carnivora. However, while becoming extinct in Eurasia during the late Oligocene, creodonts persisted and flourished in Africa, which at that time was an isolated continent. In early Miocene times, a wide array of forms were present at a number of localities of northern (Moghara, Egypt; Gebel Zelten, Libya) and eastern Africa (Rusinga, Songhor, Ombo): *Hyainailouros, Teratodon, Anasinopa, Metasinopa, Dissopsalis, Metapterodon, Leakitherium,* and *Isohyaenodon.* This array of creodonts coexisted with a number of true carnivores that entered Africa from Eurasia during the early Miocene: amphicyonids (*Amphicyon, Afrocyon, Kichenia, Cynelos*), nimravids (*Syrtomilus*), and felids (*Pseudaelurus*).

In contrast, only *Hyainailouros* entered Eurasia from Africa, accompanying the gomphotheres. The asymmetry of this carnivore–creodont dispersal is probably best explained by the differences in the locomotor abilities of the two groups. *Hyainailouros* was a giant carnivore, close to the genus *Pterodon,* with a 0.5-m-long skull and a postcranial anatomy still retaining an archaic kind of locomotion, different from that of an evolved carnivore: its forelimbs resembled more those of an ungulate than those of a modern carnivore, the pronation-supination movement being impossible. This indicates a good adaptation of *Hyainailouros* to running but not to the capture of prey. In fact, it seems probable that *Hyainailouros* was not an active predator but a scavenger, as shown by the horizontal wear of its premolars. This could explain why only two creodont forms entered Eurasia (*Hyainailouros* and *Sivapterodon*), while at least five genera of carnivores settled in Africa across the Gomphothere Bridge.

In any event, as a scavenger, *Hyainailouros* probably suffered little competition from the true hyaenids that were present in Europe. These primitive hyaenids included small insectivorous/omnivorous species of the genera *Protictitherium* and *Plioviverrops*. The former were civetlike insectivorous/omnivorous hyaenids, with generalized, civetlike dentition displaying a full set of premolars and molars. The postcranial skeleton indicates the retention of arboreal adaptations, such as retractable claws, and together with the dentition indicates a semiarboreal existence and a diet consisting of small mammals, birds, and insects, far from the lifestyle of today's hyenas. *Plioviverrops* was a mongooselike insectivorous/omnivorous carnivore that showed a progressive adaptation to insectivorism, as indicated by the reduction of the sectorial portion of the dentition and the increase in the number of high, puncturing-crushing cusps on the cheek-teeth. Its skeleton was apparently more adapted to a terrestrial lifestyle than that of *Protictitherium*. Both originated in western Europe.

Besides the large carnivores, a number of small to medium-size carnivores, such as the first genets, or viverrids (*Semigenetta elegans* at Artenay, Erkerthofen 2, Wintershof-West), and mongooses (herpestids like *Leptoplesictis aurelianensis*), were already diversified during the early Miocene in Europe (figure 4.16). But the most abundant family of small carnivores was that of the mustelids, which by the middle Miocene included most of the subfamilies we can recognize today. Starting with *Plesictis* in the early Miocene, this Oligocene survivor was in the late early Miocene replaced by a variety of forms, including several species of martens (*Martes munki* and others), wolverines and glutton-related forms (*Trochictis, Ischyrictis, Hoplictis*), badgers (*Taxodon*), skunks (*Miomephitis, Proputorius, Plesiomeles*), and otters (*Paralutra, Potamotherium*) (figures 4.17 and 4.18). At this time, the raccoons, or procyonids, of the species *Schlossericyon viverroides* were also present.

THE MIDDLE MIOCENE EVENT

A new marine transgression, known as the Langhian Transgression, characterized the beginning of the middle Miocene, affecting the circum-Mediterranean area. Consequently, the seaway to the Indo-Pacific re-opened for a short time, restoring the circumequatorial warm-water circulation. Stratigraphically, the worldwide appearance of the planktonic foraminifer *Praeorbulina* marks the base of the Langhian. This event coincided with a global warming indicated by a bipolar spreading of warm-water larger foraminiferous and calcareous nannoplankton into higher latitudes at around 17 to 16 million years ago. A uniform tropical and subtropical mollusk and foraminiferous fauna developed in all intramountain Paratethyan basins (Vienna, Pannonian, Transylvanian, and so forth) and in the Carpathian Foredeep. In the Vienna Basin, diverse subtropical biomes included the latest-occurring croco-

FIGURE 4.16 *Reconstructed life appearance of the Miocene herpestid* Leptoplesictis aurelianense

Direct fossil remains of Miocene herpestids (mongooses) consist mostly of teeth, which are similar to those of viverrids and tell us little about the body structure of the animal. But modern herpestids differ strikingly from viverrids in their foot structure, which is far more adapted to terrestrial locomotion and moderate digging. Footprints of unmistakable herpestid morphology have been found in the fossil track site of Salinas de Añana, Spain, demonstrating that as early as the early Miocene, there were mongooses with a modern foot pattern. This reconstruction is based on the measurements of the Salinas's track, which was left by a mongoose moving in a diagonal-sequence trot.

diles and palm trees. These tropical conditions became established as far north as Poland in marine coastal and open-sea waters.

After the optimal conditions of the early Miocene, the middle Miocene was a period of global oceanic reorganization, representing a major change in the climatic evolution of the Cenozoic. Before this process began, high-latitude paleoclimatic conditions were generally warm although oscillating, but they rapidly cooled thereafter, leading to an abrupt high-latitude cooling event at about 14.5 million years ago that is well recognized in the oceanic stable-isotope record (Miller et al. 1991; Flower and Kennett 1994). Increased production of cold, deep Antarctic waters caused the extinction of several oceanic benthic foraminifers that had persisted from the late Oligocene–early Miocene and promoted a significant evolutionary turnover of the oceanic assemblages from about 16 to 14 million years ago (Woodruf 1985).

This middle Miocene cooling was associated with a major growth of the Eastern Antarctic Ice Sheets (EAIS) and started the climatic regime that would characterize the late Neogene times (in particular, the period between 14.5 and 14.1 million years ago records the largest increase in semipermanent EAIS; Shackleton and Kennett 1975; Wright et al.

1992). Continental ice sheets had certainly been present since the Oligocene (Miller et al. 1987) but probably were less extensive than at present (it has been suggested that the Eastern Antarctic Ice Sheet was of temperate rather than polar climatic character). Ice-sheet fluctuations must have been greater during the interval from 16.6 to 14.9 million years ago than during the time from 14.9 to 12 million years ago. Consequently, middle Miocene sea levels experienced large, short-term fluctuations of 50 m from about 16 to 14 million years ago, including three high stands separated by two sea-level falls. Decreased variability in oxygen isotopic records from 14.5 to 14.2 million years ago suggests that the Antarctic cryospheric development resulted in an increased stability of EAIS and the corresponding decrease in the variability of the climate system. A permanent global sea-level fall at 14.2 million years

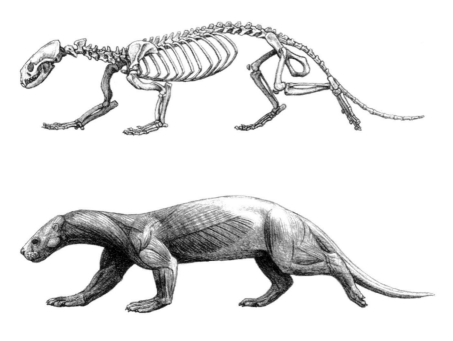

FIGURE 4.17 *Skeleton and reconstructed musculature of the mustelid* Potamotherium valletoni

Abundant and well-preserved fossils of this carnivore from the site of Saint-Gérand-le-Puy give us a clear picture of its habits and appearance. The long trunk and tail and the short, heavily muscled limbs show an adaptation to aquatic life even more extreme than that of extant river otters. Like them, *Potamotherium valletoni* would have been rather slow and clumsy when walking on land, but it would have been a graceful swimmer, thanks to the powerful thrust provided by strong limb muscles. Reconstructed head and body length: 55 cm.

FIGURE 4.18 *Reconstructed head of* Potamotherium valletoni

With a short muzzle, a long brain case, and large orbits, the head of *Potamotherium* would have looked distinctly otterlike in life. Other features for which we do not have direct evidence, such as small ears and well-developed whiskers, were probably present as well: the ears become small in all aquatic carnivores in order not to break the hydrodynamic shape of the head, while large whiskers are important for the animals to feel their way in the water. Their presence in *Potamotherium* is at least suggested by the great diameter of the infraorbital foramen, through which pass the nerves that supply the roots of the whiskers. Wide infraorbital foramina are present in all carnivorous species that have large and sensitive whiskers. Basal length of skull: 10 cm.

ago, which followed this period of instability, confirms this analysis (Haq et al. 1987).

What could have been the cause of such a change in the climatic–oceanic system during the middle Miocene? There are at least two possible causes. First, we must consider the tectonic events that affected the oceanic circulation at that time. During the interval represented by the Langhian Transgression, previous to the cooling, the widespread evaporation of the eastern Mediterranean and Paratethys Seas may have contributed to an increased production of warm, deep saline waters through the eastern Mediterranean–Indo-Pacific seaway. This Tethyan–Indian deep saline water upwelled in the Antarctic region throughout the early Miocene. Meridional transport of heat by this deep-water source may have prevented large-scale growth of the EAIS through the early Miocene to the middle Miocene, until approximately 14 million years ago (Woodruf and Savin 1989).

But following the Langhian Transgression, deep-water circulation underwent major changes. In Serravallian times, tectonic movements along the Levant and the eastern and northern Anatolian faults began. Simultaneously, the uplift of the Himalayan and Alpine systems progressed (resulting in further erosion and the sequestering of organic carbon), and the Rift Valley formed in eastern Africa (Axelrod and Raven 1978). By the mid-Serravallian, the final disconnection of the Paratethys occurred. The short-lived marine connection between the Mediterranean Sea and the Indian Ocean in the Bitlis Zone closed again, as well as the seaway to the Transcaucasian Basin. A first effect of this event was the desiccation of the Paratethys Sea (the so-called Badenian salinity crisis) and the deposition of thick evaporite deposits of gypsum and halite covering the foredeep from Romania to Poland (where the famous salt mine of Wieliczka is located). But the main effect of the closure of the communication between the western Mediterranean Sea and the Indic domain was the termination of the Tethyan–Indian deep saline water, which probably involved a decrease in poleward heat transport. In fact, the end of this Tethyan source of warm deep water for the Antarctic at approximately 14 million years ago closely coincided with the time of major EAIS growth from 14.5 to 14.1 million years ago.

A second way to explain the climatic crisis of the middle Miocene is the so-called Monterey Hypothesis, which suggests that episodic organic carbon burial accelerated global cooling from 15.6 to 15.4 million years ago and from 14.5 to 14.1 million years ago. The sequestering of large quantities of organic carbon in marginal basin sediments around the Monterey Formation of California and the phosphate deposits of the southeastern United States may have led to global cooling and EAIS growth through the consequent partial drawdown of atmospheric CO_2 (Vincent and Berger 1985). A positive correlation has been found between the proportion of carbon being stored as organic carbon in terrestrial and marine marginal basin reservoirs relative to calcium carbonate sediments in the deep sea, and the mean ocean values of the isotope $\delta^{13}C$ (Vincent and Berger 1985). Increased carbonate preservation and $\delta^{13}C$ maxima in the deep-sea record mark intervals of high organic carbon burial in marginal basin sediments. Conversely, the transfer of organic carbon to the deep ocean lowers the mean ocean $\delta^{13}C$ and raises the values of CO_2, resulting in increased dissolution and reduced carbon preservations. In this way, a broad positive $\delta^{13}C$ excursion and consequent depletion of atmospheric CO_2 has been recorded during the early to middle Miocene, coinciding with the deposition of large quantities of organic carbon in marginal basin sediments around the Monterey Formation and some phosphate deposits in the southern

United States. Besides these two areas in North America, the uplift of the Alpine and Himalayan systems may also have enhanced the sequestering and burial of large quantities of organic carbon. All these factors, including the termination of the Tethyan–Indian warm-water circulation, must have contributed together or separately to the global cooling and EAIS growth recorded from 14.5 to 14.1 million years ago.

THE MIDDLE MIOCENE EVENT AND THE FIRST DEVELOPMENT OF OPEN WOODLANDS IN EUROPE: THE GREEK-IRANIAN PROVINCE

Middle Miocene polar cooling and east Antarctic ice growth had severe effects on middle- to low-latitude terrestrial environments. There was a climatic trend to cooler winters and decreased summer rainfall. Seasonal, summer-drought-adapted schlerophyllous vegetation progressively evolved and spread geographically during the Miocene, replacing the laurophyllous evergreen forests that were adapted to moist, subtropical and tropical conditions with temperate winters and abundant summer rainfalls (Axelrod 1975). These effects were clearly seen in a wide area to the south of the Paratethys Sea, extending from eastern Europe to western Asia. According to the ideas of the American paleontologist Ray Bernor, this region, known as the Greek-Iranian (or sub-Paratethyan) Province, acted as a woodland environmental "hub" for a corridor of open habitats that extended from northwestern Africa eastward across Arabia into Afghanistan, north into the eastern Mediterranean area, and northeast into northern China. The Greek-Iranian Province records the first evidence of open woodlands in which a number of large, progressive open-country mammals—such as hyaenids, thick-enameled hominoids, bovids, and giraffids—diversified and dispersed into eastern Africa and southwestern Asia. The mammalian composition of the Greek-Iranian Province was quite different from that of the more wooded environments that persisted in most of western and central Europe and approached in some ways that of the recent African savannas. This is why it has often been regarded as a "savanna-mosaic" chronofauna. But in fact, as demonstrated by several analyses, this eastern biome was closer to an open sclerophyllous woodland than to the extensive grasslands present today in parts of Africa.

Nevertheless, the peculiar biotope developed in the Greek-Iranian Province acted as the background from which the African savannas evolved during the Pliocene and Pleistocene. They also included a number of genera in common with the open-country chronofauna that dominated the late Miocene in Eurasia. This evolution has been carefully followed in eastern Africa, where a similar ecosystem of seasonally adapted schlerophyllous woodland, with terrestrial hominoids (*Kenyapithecus*), was present in Fort Ternan (Kenya) as early as 14 million years ago, associated with the first extension of grasslands into eastern Africa.

The most outstanding effect of the Middle Miocene Event is seen among the herbivorous community, which showed a trend toward developing larger body sizes, more-hypsodont teeth, and more-elongated distal limb segments (Janis and Damuth 1990). Increasing body size in herbivores is related to a higher ingestion of fibrous and low-quality vegetation. Browsers and grazers have to be large because they need long stomachs and intestines to process a large quantity of low-energy food (this is why they have to eat almost continuously). Because of the mechanism of rumination, ruminants are the only herbivores that can escape this rule and subsist at small sizes. Increasing hypsodonty and high-crowned teeth are directly related to the ingestion of more-abrasive vegetation, such as sclerophyllous leaves or grasses, or to the development of grazing or rooting behaviors that involve a higher minerogenic component in the diet. Finally, the elongation of the distal limb segments is related to increasing cursoriality. The origin of cursoriality can be linked to the expansion of the home range in open, low-productive habitats.

At the taxonomic level, this habitat change in the low latitudes involved the rapid adaptive radiation of woodland ruminants (bovids and giraffids). More particularly, the bovid diversity suddenly increased in the Greek-Iranian Province. Although the small *Eotragus* persisted and was found throughout Europe, western Asia, and Africa, a new group of larger bovids, the boselaphines, spread at this time, becoming the most successful members of this family during most of the Miocene. The boselaphines today include only the nilgai (*Boselaphus tragocamelus*), a large horselike antelope that inhabits the high woodlands of India (its composite Latin name, which means "cow-deer," defines its morphology quite well).

But in the middle Miocene, this group began a successful evolutionary radiation that led to a large generic diversity. *Miotragocerus,* the first boselaphine appearing in Europe, is recorded in Georgia (Belometchetskaya), Turkey (Eçme-Akçaköy), and some middle Miocene faunas from Spain (Tarazona). It was a medium-size bovid of about 80 kg, with strong horn cores that looked very different from those of *Eotragus*. Its teeth were still primitive but with some cement. The limb bones and foot anatomy indicate that *Miotragocerus* lived in humid habitats where it probably fed on soft plants (figure 5.13).

At that time, a second boselaphine bovid in the Greek-Iranian Province, *Austroportax,* appeared. *Austroportax* was a large and surprisingly advanced bovid for its time. It weighed about 300 kg and was supported by short, heavy extremities resembling those of today's buffaloes and other modern members of the Bovini tribe. *Austroportax*'s foot morphology indicates that it lived in humid and wooded habitats. Like that of *Miotragocerus*, the dentition of *Austroportax* suggests a diet of soft vegetation.

The genera *Tethytragus* and *Gentrytragus* represented a second wave of successful bovids that spread throughout southern Europe and Africa. Both originated from an Asian form called *Caprotragoides*, which was first found in older middle Miocene deposits of Siwaliks, Pakistan. Following the environmental changes in the middle Miocene, *Caprotragoides* spread over Europe and northern Africa throughout the Greek-Iranian Province, leading to *Tethytragus* and *Gentrytragus*, respectively. All were medium-size bovids of about 30 kg with horn cores curved backward and slightly outward. The dental morphology indicates a diet based on a great variety of plants. The limb proportions resemble those of the reedbuck (*Redunca redunca*). Most of the characteristics of these three genera seem to be adaptations to dwelling in open country, but other traits indicate more wooded preferences or suggest rolling to mountainous country. This mosaic of features indicates that *Caprotragoides*, *Tethytragus*, and *Gentrytragus* were probably ubiquitous bovids with a high capability to invade diverse biotopes.

A third group of advanced bovids that spread at this time were the hypsodontines. *Hypsodontus* was a medium to large (about 110 kg), slender, and specialized long-legged bovid. It differed from *Eotragus* and *Tethytragus* in its early acquisition of hypsodont cheek-teeth and in the torsion of its horn cores. The extremely hypsodont cheek-teeth indicate a diet based on grass and tough plants. These animals attained a broad Eurasian distribution during the middle Miocene, from China to India, Georgia (*Hypsodontus miocaenicus* at Belometchetskaya), eastern Europe (*Hypsodontus serbicum* in Yugoslavia), and Africa. A second genus related to *Hypsodontus*, *Turcocerus*, was present in Turkey at the same time. *Turcocerus* was a small bovid with slender although massive metapodials. It bore two short conical horns showing clockwise torsion. The teeth were also extremely hypsodont and with cement, indicating a diet based on leaves, herbs, and grasses.

A fourth group of advanced bovids to spread during the middle Miocene were the antilopines, represented mainly by the gazelles. The first gazelles (*Gazella*) came from the early Miocene beds of the Chinji zone of Siwaliks and from Majiwa, Kenya (Thomas 1984). According to these data, *Gazella* and other antilopines could have originated in Africa or the Siwaliks from a form close to *Homoiodorcas* or a related neotragine. Gazelles dispersed into Europe at this time from their possible Afro-Arabian origins, perhaps taking part in the same dispersal event as *Giraffokeryx* and the kubanochoers.

Not only gazelles but also the giraffids experienced a wide adaptive radiation into Africa after their dispersal from Asia. This evolutionary radiation paralleled in some ways that of the paleomerycids into Europe, leading to a variety of forms with sophisticated cranial appendages (as in the case of *Prolibytherium,* which bore a pair of fan-shaped ossicones). One of these forms, *Giraffokeryx,* dispersed out of Africa and became widespread at this time, remains having been found at several middle Miocene localities of the Greek-Iranian Province, such as Chios, Paçalar, Prebreza, and Belometchetskaya, as well as in the Bugti Beds, Pakistan. *Giraffokeryx* displayed two pairs of rather short, unbranched ossicones. The anterior pair was situated in front of the orbits, while the second pair rose directly behind the orbits. A second giraffid, *Georgiomeryx,* has also been found in some middle Miocene localities of the Greek-Iranian Province, like Chios in Greece. *Georgiomeryx* was closely related to *Giraffokeryx* but displayed a single pair of flat, laterally extended ossicones over the orbits and had a more archaic dentition with brachydont teeth.

In contrast with the highly diversified bovid fauna, the cervoid representation in the open woodland areas of the Greek-Iranian Province was extremely poor, reduced almost exclusively to primitive moschids of the genera *Hispanomeryx* and *Micromeryx.* As with today's representatives of this family, both genera were hornless ruminants displaying, in the males, prominent canines. *Hispanomeryx* was a small animal (about 5 kg) with a dentition adapted to foraging in the undergrowth in search of soft plants, leaves, larvae, and, from time to time, even carrion. It has been found in a number of middle Miocene Greek and Turkish localities (Chios, Çandir, and possibly Paçalar), but its range can be extended to central Spain. *Micromeryx* was a small and gracile moschid of less than 5 kg (figure 4.19). Its dentition was similar to that of the duikers (*Cephalophus*) but more primitive. Like *Hispanomeryx, Micromeryx* probably foraged in the lower vegetation of the closed forest, living on soft plants and fruits, larvae, and carrion. The geographic range of *Micromeryx* was similar to that of *Hispanomeryx,* ranging from Turkey and other places of the Greek-Iranian Province to central Spain. However, while *Hispanomeryx* was a short-lived form that disappeared during the middle Miocene, *Micromeryx* was a successful moschid that spread over a wide area of western and central Europe and persisted until the late Miocene (early Turolian).

Among the suids, the listriodontines evolved in a peculiar way in northern Africa, leading to giant forms such as *Kubanochoerus,* with a weight of about 500 kg, which in some species may have reached 800 kg. *Kubanochoerus* was found for the first time in Belometchetskaya, Georgia, and probably derives from the African *Libyochoerus,* from the early Miocene locality of Gebel Zelten (figure 4.20). The most striking feature of

FIGURE 4.19 *Reconstructed life appearance of the moschid* Micromeryx

Micromeryx was a tiny animal, broadly similar to modern musk deer but somewhat smaller. The tusks that protruded beyond the upper lips would have been used for intraspecific fights, as with modern moschids, and their fearsome appearance contrasted with the elegant, almost fragile build of this small ruminant.

these giant listriodontines was the presence in the males of an enormous horn above the orbits, which was probably used for intraspecific fighting and which indicates a unique case of territoriality among suids. Together with the listriodontines, a new subfamily of suids, the tetraconodontines (*Conohyus, Parachleuastochoerus, Sivachoerus, Notochoerus,* and the African *Nyanzachoerus*) became the dominant suiforms in the circum-Mediterranean area. The tetraconodonts bore thick enameled cheek-teeth and conical premolars with hyena-like wear, which probably indicates a diet based on such hard food items as seeds. A trend toward reducing size was present in this group during the middle Miocene, from the 70 kg of the medium-size *Conohyus* to the 40 kg of the small *Parachleuastochoerus.* They replaced the former, early Miocene–dominant, generalized hyotherine suids, such as *Aureliachoerus* and *Hyotherium.*

While the tetraconodonts such as *Conohyus* were dominant in western Europe, the paleochoerids persisted and succeeded in the Greek-

FIGURE 4.20 *Two views of the skull and reconstructed head of the suid* Kubanochoerus gigas

This reconstruction is based on a skull from Belometchetskaya, Georgia, with additional information from skulls from Miocene sites in China. The frontal horn was probably used during fights between males of the species, suggesting a combat style vaguely similar to that of the modern giant forest hogs (*Hylochoerus*) from Africa, which hit each other violently in the forehead and have correspondingly thickened frontals.

Iranian Province by developing peculiar feeding adaptations. This was the case with the sanitheres, a group of small suids with selenodont cheek-teeth. Their molarized premolars and molars bore a wrinkled enamel formed of several cuspules and ridges, well adapted for browsing leaves and other sclerophyllous vegetation. A second browsing pig was *Schizochoerus*, a small peccary-like suid related to *Taucanamo* that developed lophodont molars and short, broad premolars. This dentition resembled that of the contemporaneous advanced listriodontines and, as in that group, was probably well adapted for browsing the tough vegetation of the sclerophyllous evergreen woodland that covered most of the Greek-Iranian Province at that time.

Among the large browsers, the aceratherine rhinos persisted in the eastern province but had to coexist with some Asian immigrants, such as the elasmotherine rhinos of the genus *Hispanotherium* (= *Begertherium* from Paçalar; Fortelius 1990). In contrast to its early Miocene semiaquatic relatives, such as *Brachypotherium*, *Hispanotherium* was much better adapted to the conditions of seasonal dryness and sclerophyllous vegetation dominant at that time, with long and slender limb bones and hypsodont cheek-teeth (plate 7). Moreover, its molars showed a complicated design, with a number of reentrant ridges filled with dental cement. The dental wear of later forms indicates that they were specialized grazers, as confirmed by the adult wear-facet development in *Hispanotherium* from Paçalar, which was approximately similar to that of the grazer *Rhinoceros unicornis*.

Hispanotherium raises a peculiar zoogeographic problem because, as indicated by its generic name, the first remains of these elasmotherines were not described in Turkey but in the middle Miocene sediments of central Spain. However, with the exception of some scarce remains in southern France, this genus is absent from the rich middle Miocene record of western and central Europe and from Africa. It has been suggested that during the middle Miocene a system of land bridges directly connected the Iberian Peninsula with Anatolia and the Greek-Iranian Province throughout the Mediterranean; this could explain the near absence of *Hispanotherium* faunas in other parts of central and western Europe. These middle Miocene faunas from central Spain (several localities in the Calatayud-Daroca, Tagus, Madrid, and Duero Basins) are very peculiar and differ from a paleoecological point of view from those of western Europe. Browsers such as the first boselaphines of the genus *Miotragocerus* (Tarazona), the tetraconodontine suids of the genus *Conohyus*, and some moschid cervoids (*Micromeryx*) dominated the large-mammal faunas. At this time, the highly diversified early Miocene suid representation also dropped, and the listriodontine suids of the genus *Bunolistriodon* became dominant. From the rich cervid fauna of the early Miocene, only *Procervulus* persisted, the cervids of the genus

Lagomeryx becoming extinct in this area. In contrast, the palaeomery-cids became the dominant ruminants, represented by the species *Triceromeryx pachecoi* (plate 7).

The small-mammal faunas also were affected by the drier conditions. The opossum-like marsupials like *Amphiperatherium* finally disappeared from Europe, after millions of years of successful persistence (since the late Cretaceous). A clear drop in diversity also affected most of the species living in closed forests, such as dormice (glirids) and gliding rodents (flying squirrels and eomyids). A group of ubiquitous cricetids, those of the genus *Megacricetodon*, became dominant among the rodent faunas, with a vast array of vicarious species in each region. In Iberia, the glirids of the genus *Armantomys* developed a unique dental adaptation as a response to the dominant sclerophyllous conditions. Their cheek-teeth became increasingly hypsodont, while their general design became simpler, finally reduced to three main transverse ridges. This kind of evolution occurred but rarely among dormice, which did not tend to develop high-crowned, hypsodont cheek-teeth. Among the squirrels, the ground squirrels of the tribe Xerini, such as *Atlantoxerus*, became the most frequent. Today's representatives of *Atlantoxerus* live in the desert and subdesert landscapes of northern Africa.

Like *Armantomys*, other rodent families developed dental adaptions in response to the dominant sclerophyllous vegetation. For instance, a group of cricetid rodents exhibited a trend toward developing selenodont, ruminant-like cheek-teeth. The first member of this group, *Cricetodon*, was a generalized hamster with a primitive dental design. However, its descendants gradually developed longitudinal ridges (called ectolophes) connecting the labial cups of these teeth and leading to a selenodont pattern similar to that of the ruminants. This trend also occurred in a number of middle to late Miocene genera, such as *Hispanomys* in western Europe, *Deperetomys* in central Europe, and *Byzantinia* in eastern Europe. At the same time, in most of these lineages ectolophes developed in parallel with increasing hypsodonty, which reached a maximum in some Pliocene survivors, such as *Ruscinomys*. In fact, the first remains of *Byzantinia*, found in the Greek locality of Samos, were initially assigned to a "small ruminant," their real cricetid affinities not being determined until years later. Increasing hypsodonty and ectoloph development appeared independently in such other cricetid lineages as the myocricetodontines, a group of Afro-Asian cricetids. The genus *Zramys*, from the late Miocene in northern Africa, can hardly be distinguished from the European representatives of *Hispanomys* and *Ruscinomys*.

Among the carnivores, one of the families most affected by the Middle Miocene Event was the amphicyonids. This group declined in diversity, with a number of species disappearing during the middle Mio-

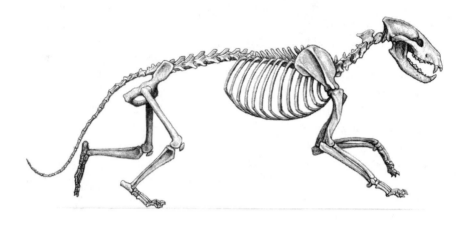

FIGURE 4.21 *Skeleton and reconstructed life appearance of the amphicyonid* Amphicyon major

Abundant fossils of this carnivore, including a nearly complete skeleton, have been found at Sansan. The body proportions of this species were a strange mixture of bearlike and catlike features. Like bears, *Amphicyon* had broad feet with splaying digits and long, nonretractile claws, and the posture was probably plantigrade or semiplantigrade. But it resembled modern big cats in having a long and flexible back and tail, a long and narrow ilium, and hind limb bones designed for powerful jumping. Such proportions fit well with ambush hunting and suggest that active predation was a significant part of its feeding behavior. With its vaguely doglike skull, catlike body, and bearlike feet, *Amphicyon* would not have looked like any modern big carnivore.

cene: *Cynelos schlosseri* and *C. helbingi* (last recorded at Esvres and Pontlevoy, France), *Ictiocyon socialis, Ysengrinia valentiana, Euroamphicyon olisiponensis,* and *Amphicyon giganteus.* A first accounting of the variety of morphotypes represented by these losses gives no clear pattern to these extinctions. Some of the extinct species were small omnivorous or mesocarnivorous forms, such as *I. socialis* and *C. helbingi.* However, *E. olisiponensis* and *A. giganteus* were large, bone-crushing hunters. And *Y. valentiana* was a small hypercarnivore. It seems that the middle Miocene climatic cooling affected the small generalized omnivorous and mesocarnivorous amphicyonids in a quite specific way, while the large bone-crushing mesocarnivorous and the hypercarnivorous morphotypes persisted, with new amphicyonid species replacing the losses among these last morphotypes.

This period records the first appearance of one of the most successful amphicyonid species in the Miocene, *Amphicyon major* (starting from Neudorf-Spalte and several localities in Spain, France, Germany, the Czech Republic, and Turkey), which replaced the rather similar *E. olisiponensis* and *A. giganteus.* Males of *A. major* attained 212 kg and females 122 kg, the size of a lion but far away from the enormous dimensions of the early Miocene *A. giganteus* (figure 4.21). But *A. major* displayed body proportions and teeth adaptations similar to those of *A. giganteus.* *A. major* had a long skull (relatively longer than those of today's canids, ursids, mustelids, and felids), with an elongated snout and relatively long and massive canines. Like other amphicyon species, *A. major* possessed rather complete dentition, with four premolars and three molars (which is an archaic feature). The general limb proportions were similar to those of a bear, with short metapodials. All these features indicate that *A. major* was an active hunter with good capabilities for bone-crushing.

Among the hypercarnivores, the entry of *Thaumastocyon,* a medium-size amphicyonid of about 85 kg that was probably related to *Agnotherium grivense,* compensated for the loss of *Ysengrinia. Thaumastocyon dirus* was a bizarre hypercarnivore that had lost its last molars (a peculiar case among amphicyonids). Like *Agnotherium, T. dirus* had a body plan clearly more cursorial than that of other members of the group and probably hunted by pursuit. It is first know from the middle Miocene locality of Pontlevoy.

The middle Miocene also marked the first appearance of the hyaenids (*Protictitherium*) in eastern Europe and western Asia (Greek-Iranian Province). But as noted earlier, these primitive hyaenids were more-generalized civetlike carnivores than true scavengers. This does not mean that the hyena-like scavenger guild was empty at that time, since a peculiar family of carnivores, the percrocutids, inhabited the Greek-Iranian Province. The members of this family (*Percrocuta, Dinocrocuta,*

Allohyaena) were once identified as true hyaenids related to the large scavenger hyaenids of the late Miocene like *Adcrocuta*. However, differences in the deciduous dentition demonstrated that they were part of a different family that probably developed hyena-like dental adaptations simultaneously with the true hyaenids. The percrocutids seem to correspond to an early feloid radiation covering the "hyena guild," at a time when the true hyaenids (*Protictitherium*) had not yet developed the dental and locomotor adaptations to scavenging and bone-cracking characteristic of the later members of the family. The first percrocutids belonged to the genus *Percrocuta* and were found in late middle Miocene localities of western Europe (Sansan, La Grive) and western Asia (Çandir, Paçalar), where they tended to coexist with the small, arboreal, primitive hyaenids of the genera *Protictitherium* and *Plioviverrops*. According to the fossil record, the percrocutids appear to have been a mainly Asian group that expanded successfully into the open-woodland environmental conditions of the Greek-Iranian Province. From there, they intermittently colonized western Europe, although the record of this group in this area is much more restricted. The percrocutids were generalized scavengers with a tendency toward enlarged premolars and increasing body size. However, the middle Miocene *Percrocuta* had not yet developed the bone-cracking adaptations that would be common among the late Miocene members of the family (*Dinocrocuta*).

MIDDLE MIOCENE PRIMATE DISPERSALS

The primates, which disappeared from Europe after the "Grande Coupure," pursued their evolution in Africa, where the first anthropoids, or simians, are recorded as early as 45 million years ago (*Algeripithecus minutus*, from the early middle Eocene of Glib Zegdou, Algeria; Godinot and Mahboubi 1992). They soon evolved into a variety of genera, such as *Aegyptopithecus, Propliopithecus, Apidium, Parapithecus,* and other well-known forms from the late Eocene beds of El Fayoum. Although the first evidence of African faunas in Europe after the establishment of the Gomphothere Bridge can be traced back to 19 million years ago, these primates did not enter Europe until the middle Miocene, about 16 million years ago. The small anthropoid *Pliopithecus*, dating from this time, is found in a number of localities in the Loire Valley, France (Faluns de Touraine, Anjou, Pontlevoy-Thenay, and others). *Pliopithecus*, discovered by Edouard Lartet in 1834 in the hill of Sansan (Gers), was the first fossil primate to be described in the history of paleontology. During later decades, it was considered a small hylobatid (the group of gibbons and siamangs), but we know today that it was a member of the same evolutionary radiation that led to *Aegyptopithecus, Propliopithecus, Oligopithecus,* and other archaic catarrhines from El Fayoum.

The pliopithecids were small anthropoids weighing no more than 10 kg. They probably lacked a tail, and their dentition indicates a mainly folivorous diet. The best-preserved remains belong to the species *Pliopithecus vindoboniensis*, from the middle Miocene of Neudorf an der Marche, Slovakia. This locality delivered several cranial and postcranial remains that established an accurate idea of this species's anatomy. Its face was short and wide, with large, circular orbits placed in an anterior position. This morphology contrasts with the elongated muzzle of *Aegyptopithecus* and other archaic catarrhines and resembles that of today's gibbons (although, as we have seen, the pliopithecids were not directly related to these lesser apes). The limbs were relatively long and slender, suggesting that *Pliopithecus* was an arboreal primate that, like the gibbons, performed suspensory locomotion from tree to tree. The limb proportions, however, were quite different from those of gibbons and closer to those of a basal primate, with arms and legs of similar length. As with several modern primates, *Pliopithecus* had a marked sexual dimorphism, consisting of the development in the males of sagittal cranial crests and of canines bigger than those in the females.

After their dispersal out of Africa, the pliopithecids attained a wide geographic range, from the Iberian Peninsula (Vallès-Penedès Basin) to China, throughout France, Switzerland, Germany, Poland, Austria, and Slovakia. At the same time that they expanded their range, their taxonomic diversity also increased, reaching up to five genera and more than ten species. The new genera that appeared during the middle and late Miocene, like *Plesiopliopithecus* and *Anapithecus*, emphasized the folivorous adaptations already present in *Pliopithecus*.

Besides the pliopithecids, *Griphopithecus*, a true hominoid, entered the Greek-Iranian Province from Africa between 16 and 14 million years ago. We know *Griphopithecus* mainly by dental and mandibular remains in a number of eastern European localities, such as Klein-Hadersdorf (Austria), Neudorf-Sandberg (Slovakia), and Çandir and Paçalar (Turkey). This hominoid was, in fact, a direct descendant of (and perhaps synonymous with) *Kenyapithecus*, a new type of anthropoid that appeared in eastern Africa during the middle Miocene. *Kenyapithecus* differed in a number of features from the early Miocene proconsulids that descended from the primitive catarrhines like *Aegyptopithecus* in Africa. Unlike the arboreal *Proconsul*, *Kenyapithecus* was a semiterrestrial quadruped that probably spent most of its time on the forest ground (although it could quickly climb trees in case of danger). A thick enamel, much different from that of *Proconsul* and the pliopithecids, characterized the dentition of *Kenyapithecus*, which was adapted to a diet based on seeds and tough vegetables. This adaptation coincided with the development at about 14 million years ago of sclerophyllous vegetation in extensive areas of eastern Africa and the Greek-Iranian Province.

The Latest Middle Miocene Faunas

Despite the effects of the Middle Miocene Event on the low-latitude terrestrial ecosystems of Europe, the western and central European faunas remained almost unaffected, retaining an association of mammals that was similar to or even richer than that of the early Miocene (plate 8). In western and central Europe, the suids attained their highest diversity during the later part of the middle Miocene, with a combination of hyotherines (*Hyotherium*), peccary-like suids (*Taucanamo, Albanohyus*), tetraconodontines (*Conohyus* and its offshoot, *Parachleuastochoerus*), and listriodontines (*Listriodon*). *Albanohyus*, a small peccary-like suid also found at Fort Ternan, resembled *Taucanamo* but with smaller and shorter premolars. A major event at this time was the dispersal of the listriodontine *Listriodon*, which replaced the semilophodont members of the genus *Bunolistriodon* (plate 8). *Listriodon* was a full-lophodont, browsing listriodontine, which dispersed over Eurasia from China to the Iberian Peninsula throughout the Greek-Iranian Province and central Europe. Analysis of microwear in *Listriodon* has shown a more uniform diet with a smaller minerogenic component. This implies a variation from the typical rooting behavior of generalized suids and a specialization in browsing tough vegetation. The lengthening of distal limb segments might indicate that this listriodontine settled in more open habitats. Together with the peccary-like *Schizochoerus*, *Listriodon* was almost the only suid present in the Greek-Iranian Province.

At the end of the middle Miocene, suid diversity increased again with the appearance of new tetraconodontines (*Parachleuastochoerus*, a small descendant of *Conohyus* with narrower cheek-teeth and more reduced premolars) and the first representatives of modern suids (*Propotamochoerus*). *Propotamochoerus* was a large suid (about 120 kg) that probably arose in southern Asia from a hyotherine pig during the middle Miocene and subsequently extended its range westward into southwestern Asia and Europe. It is the first recognizable member of all modern suines, such as *Potamochoerus* (bush pig), *Hylochoerus* (giant forest hog), *Phacochoerus* (warthog), and *Sus* (wild boar and its domestic relatives). The molars of this group trend toward the proliferation of several minor cusps, concomitant with the loss of cusp identity. This peculiar dental evolution resembles that of some bears and indicates a further adaptation to omnivorism.

Among the bovids, a second boselaphine, *Protragocerus*, similar in many ways to *Miotragocerus*, appeared. In contrast to the diversity of bovids in the Greek-Iranian Province, bovid diversity continued to be low in western Europe and decreased again when the long-lived *Eotragus* and the immigrant *Tethytragus* became extinct.

The cervids maintained their high diversity in this part of western Europe. Although several archaic artiodactyl browsers, such as *Cainotherium,* the moschoid *Hispanomeryx,* and the old lagomerycids like *Lagomerx,* disappeared during the middle Miocene, the entry of such new cervids as *Heteroprox, Dicrocerus,* and *Euprox* compensated for these losses (figure 4.22). *Heteroprox* was a medium-size, long-legged cervid of about 35 kg displaying a two-pronged antler. According to its limb proportions, it was probably a semiaquatic browser living in a humid habitat. *Dicrocerus* was also a two-pronged deer with low-crowned (brachydont) dentition. However, this still-primitive deer displayed protoantlers with a burrlike, enlarged area at their base. Another difference between *Dicrocerus* and early Miocene cervids like *Lagomeryx* and *Procervulus* was that *Dicrocerus*'s antler was longer than the pedicle, about 20 cm long. A third middle Miocene cervid, *Euprox,* was close to *Heteroprox* and probably lived in closed forests under humid conditions as well. Although the antlers of *Euprox* were two-pronged, there was a principal, posterior prong and a secondary, smaller anterior prong. Nevertheless, the most outstanding difference was the presence in *Euprox* of a real burr, indicating for the first time the border between the deciduous and permanent segments of the antlers.

Among the perissodactyls, some early Miocene holdovers such as the semiamphibious teleoceratine rhino *Prosantorhinus* became extinct. However, *Brachypotherium,* a descendant of *Diaceratherium,* was widely present in western and central Europe. Like its early Miocene relative,

FIGURE 4.22 *Reconstructed life appearance of middle Miocene cervids Left to right: Paleoplayticeros, Stehlinoceros, and Heteroprox.*

Brachypotherium was a large teleoceratine with hypsodont teeth, short legs, and hippolike body proportions. Among the acerathere rhinos, the primitive *Plesiaceratherium* also became extinct, the appearance of more-derived forms like *Hoploaceratherium* and *Alicornops* largely compensating for this loss. Like its close relative *Plesiaceratherium*, *Hoploaceratherium tetradactylum* was a medium-size acerathere with long limbs and slender body proportions (plate 8). The upper incisors were absent, although the rather retracted nasals may have supported a small horn. In contrast, *Alicornops simorrense* was a small acerathere with short, tridactyl legs and strongly curved lower incisors (although, as with *Hoploaceratherium*, a small horn probably was present).

The rhinocerotines, or "modern horned rhinos," also significantly helped enlarge rhino diversity during the middle Miocene. Two new browsing rhinos, *Lartetotherium sansaniense* and *"Dicerorhinus" steinheimensis*, joined this already highly diversified group in western Europe. *Lartetotherium* was an old member of the group of "modern" horned rhinos that today occupy eastern Asia and Africa. These Miocene rhinocerotines seem closely related to today's close-to-extinct *Dicerorhinus sumatrensis*, the Sumatran rhino, and for many years were included in the same genus. *L. sansaniense* was a cursorial rhino with a single long horn. According to its rather brachydont teeth, its diet must have contained a higher quantity of soft plants and a lower proportion of the wooden parts of shrubs than that of the acerathere rhinos.

Another perissodactyl group, the chalicotheres, experienced a significant turnover during the middle Miocene. The old schizotherine chalicotheres were replaced by larger and more-derived forms such as *Chalicotherium*, which extended the browsing capabilities of their early Miocene ancestors. *Chalicotherium* bore the same features that were common among the early Miocene chalicotheres, such as clawed extremities and long forelimbs (figures 4.23 and 4.24; plate 8). However, they were bigger and had an even more derived limb structure. Their forelimbs were much longer than their hind limbs and were no longer used for conventional locomotion. These large perissodactyls were probably capable of a gorilla-like posture, standing over their hind limbs and using their forelimbs as "hands" to reach the higher vegetational levels. Unlike the Oligocene and early Miocene chalicotheres, *Chalicotherium* possessed a more brachydont and less specialized dentition, probably adapted to softer vegetation.

At the opposite extreme, an extensive variety of rodent species represented the small-herbivore guild. The late middle Miocene rodent faunas from western Europe reproduced the pattern observed during the early Miocene, with even higher diversity levels of cricetids, glirids, eomyids, sciurids, and castorids. A variety of gliding rodents, such as flying squirrels of different sizes (from the large *Albanensia* to the small

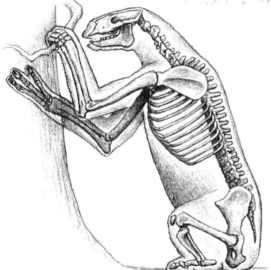

FIGURE 4.23 *Skeleton and reconstructed life appearance of the chalicotherid* Chalicotherium grande

The remains of several individuals found at the site of Neudorf, Slovakia, provide a detailed picture of this bizarre perissodactyl. The forelimbs were long and well adapted to be used as hooks to pull tree branches within reach of the mouth, while the shortened hind limbs were well suited for keeping the body in static, upright postures during long periods of browsing. When the animal reached for high branches with the forelimbs, the hind legs would be extended, with the trunk upright, to increase its effective height. Later, with the leaves already within reach of the mouth, the animal would sit, and the shortened feet and well-developed ischiadic tubers would provide a stable support for the weight of the body.

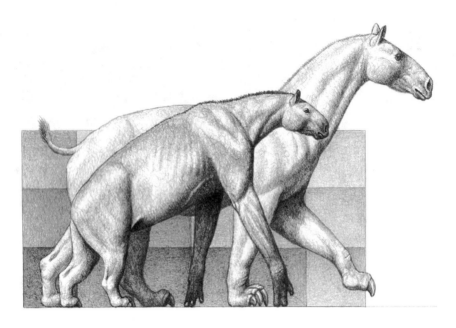

FIGURE 4.24 *Reconstructed life appearance of the Miocene chalicotherids* Chalicotherium grande *and* Ancylotherium pentelecicum, *drawn to scale*

Although these species were not strictly contemporary, they are shown side by side because they are both well known anatomically and perfectly exemplify the contrast between the two chalicotherid subfamilies. While the chalicotherines developed bizarre, vaguely gorilla-like proportions as an adaptation to a largely static style of feeding—probably in closed environments—the schizotherines, such as *Ancylotherium,* in the background, retained more "conventional" perissodactyl proportions and a more efficient quadrupedal locomotion, although in both subfamilies the feet were clawed instead of hoofed, and the forelimbs were likely used to manipulate branches and to bring the animals' forequarters up toward the lower foliage of trees. To keep such enormous claws from undue wear and to allow unhindered locomotion on firm ground, the schizotherines developed a mechanism of claw retraction that was totally different in detail from that of cats and other carnivores with retractable claws.

Blackia) and, probably, the eomyids *Eomyops* and *Keramidomys,* populated the laurophyllous forests at this time. But other groups of small rodents, such as some glirids of the genus *Glirulus,* also developed "flying" forms endowed with a patagium. This feature is quite unusual among dormice and provides strong evidence of the persistence of closed forest during the late middle Miocene in western and central Europe. Another indication of the persistence of humid conditions in this area is the frequent discovery of beavers associated with these faunas. One of them, *Chalicomys,* strongly resembled in size and morphology the recent *Castor fiber* and, like today's form, was probably highly dependent on permanent river waters. A smaller beaver, *Euroxenomys,* was widely distributed at this time and was probably associated with more unstable environments.

Among the carnivore guild, there were few changes, nimravids and amphicyonids persisting as the dominant groups. Among the nimravids, *Prosansanosmilus peregrinus* was replaced by a larger and more-derived species, *Sansanosmilus palmidens,* which in the latest middle Miocene times was itself replaced by the largest nimravid species in Europe, *S. jourdani,* of about 80 kg (figures 4.25 and 4.26).

The felids still included species of the genus *Pseudaelurus: P. turnauensis, P. lorteti,* and *P. quadridentatus.* Interestingly, *P. turnauensis* rarely appears associated with the other two species during the middle Miocene, suggesting different ecological requirements (which would explain its persistence until the late Miocene).

Among the true ursids, the mesocarnivores *Hemicyon* and *Plithocyon* and the omnivore *Ursavus* persisted. This last genus of small ursids included a larger and more-derived species, *U. primaevus,* of about 90 kg, which for some time coexisted with *U. brevirhinus.*

Among the hyaenids, a new slender form, *Thalassictis,* joined *Protictitherium* and *Plioviverrops. Thalassictis,* represented by the species *T. montadai* and *T. robusta,* was the first member of a hyaenid lineage characterized by wolflike meat and bone dietary habits. They were larger than *Protictitherium* (between 20 and 30 kg) and had less-developed posterior molars. They retained an unspecialized dentition, in some ways similar to that of canids, although with a major emphasis on bone-eating. The postcranial skeleton was adapted to terrestrial locomotion but with only moderate adaptations for cursoriality. Unlike *Protictitherium* and *Plioviverrops, Thalassictis* was adapted to a more open woodland environment. *Thalassictis,* as *Protictitherium* and *Plioviverrops* did previously, probably originated in western Europe and later dispersed into the Greek-Iranian Province.

Finally, at the very end of the middle Miocene, a number of animals of probable African origin entered Eurasia, such as the hominoids of the genus *Dryopithecus,* the giraffids of the genus *Palaeotragus,* and, perhaps, the mastodons of the genus *Tetralophodon,* which replaced *Gomphotherium. Tetralophodon* was larger and more hypsodont than *Gomphotherium.* Moreover, its skull, although still bearing four tusks, was shorter and more elephant-like, with a pair of long and straight tusks in the upper maxilla and a small pair of tusks at the end of the mandible. The first true giraffids of the genus *Palaeotragus* also entered at the end of this period, replacing the last *Giraffokeryx* in the Greek-Iranian Province. *Palaeotragus* was a relatively small giraffid of about 250 kg, bearing a pair of parallel ossicones standing upright over the orbits. It was a slender form, with long legs and limb proportions resembling those of today's

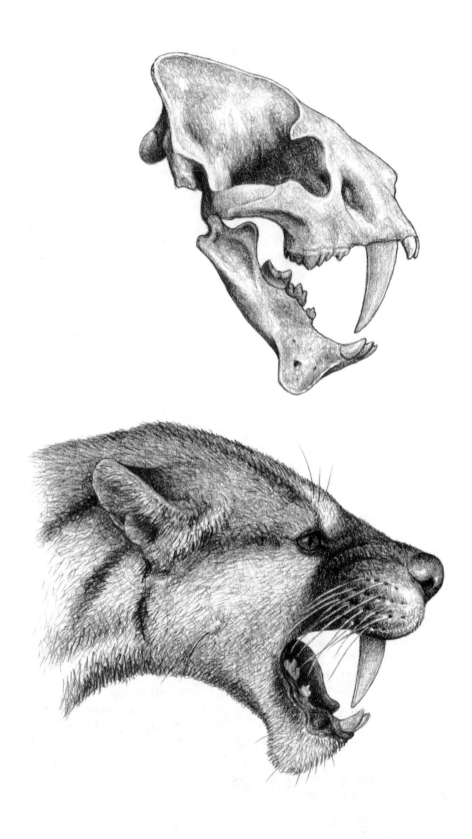

okapi. The structure of the foot indicates that *Palaeotragus* was probably an open-country runner and, perhaps, a good jumper. It probably ate soft plants, mainly leaves, which it took by grasping with a long tongue. Less common than the previous taxa was another genus of African origin, the rhinoceros *Diceros* (represented today by the black rhino, *D. bicornis*), which was found in some localities of the eastern Mediterranean. But, certainly, one of the most interesting groups of this late Aragonian African dispersal was the hominoids of the genus *Dryopithecus*.

Dryopithecus AND THE ORIGINS OF THE GREAT APE CLADE

The remains of the fossil hominoid *Dryopithecus* are well known from several middle (Aragonian) and early late Miocene (Vallesian) localities in Europe. The oldest remains correspond to isolated teeth found in the late Aragonian sites of Sant Quirze, in the Vallès-Penedès Basin (Spain), and La Grive M (France), its range extending during the early Vallesian to other localities of western and eastern Europe, like Saint Gaudans (France) and El Firal (Spain), in the Pyrenees (the localities where the type-species *Dryopithecus fontani* was first described), and Rudabanya (Hungary), in the Carpathian Basin. The last record of these hominoids occurs in the early Turolian sequence of Udabno, in Georgia (described there as *Udabnopithecus*). Unlike those of *Kenyapithecus, Griphopithecus,* and later Miocene hominoids like *Ankarapithecus, Ouranopithecus,* and *Sivapithecus,* the teeth of *Dryopithecus* were thin-enameled, indicating a different diet based on soft fruits. Until recently, little else was known about the biology of this hominoid, but the discovery of a

....

FIGURE 4.25 *Skull and reconstructed head of the nimravid* Sansanosmilus palmidens

The best fossils of this taxon come from Sansan—hence the generic name—and include several partial skulls and mandibles that have been combined to produce this reconstruction. *Sansanosmilus palmidens* was one of the earlier, but already highly specialized, members of the barbourofeline tribe. Besides the elongation and lateral flattening of the upper canines, it displayed other advanced features, such as an enlarged and downward-projecting mastoid process, a reduced coronoid process in the mandible, and a large mandibular flange. All of these were adaptations to biting with the jaws wide open, which reduces the leverage of jaw-closing muscles and makes it necessary to recruit the anterior neck muscles to depress the head and help sink the upper canines into the prey's flesh. The number of cheek-teeth was reduced, and the carnassials had huge shearing blades—features betraying a diet consisting exclusively of meat. Although *Sansanosmilus* was no more closely related to modern cats than is a civet or a hyena, the head of the living animal would have looked somewhat catlike because of the proportions of the short-muzzled skull.

FIGURE 4.26 *A scene from the site of Sansan, France, with the nimravid* Sansanosmilus palmidens *attacking an early deer,* Heteroprox larteti

The skeleton of *Sansanosmilus* was quite robust for an animal its size (smaller than the average leopard) and shows adaptations for a hunting strategy that was probably similar in most saber-toothed predators, from creodonts and marsupials to felids and nimravids. First the prey animal was subdued and immobilized with the strong forearms, and only then were the upper canines used to produce a deadly wound in the throat, using a specialized biting technique called the canine shear-bite.

rather complete skeleton in the Vallesian site of Can Llobateres in 1996 shed considerable light on this segment of primate evolution.

Between the early Miocene quadrupedal proconsulid *Proconsul* and the early Pliocene australopithecid *Australopithecus afarensis* (less than 4 million years old), little was known about the locomotion of the apes inhabiting Africa during most of the Miocene. The discovery of Can Llobateres filled this "black hole" of hominid evolution and opened a window into the world of the Miocene hominoids of Eurasia. Besides a fragmentary skull, the specimen included several vertebrae and ribs, two femurs, a tibia, and, what is most important, a rather complete arm, including the hand. The *Dryopithecus* hand was one of the most surprising elements, showing extraordinarily enlarged metacarpals and phalanges. This structure indicates a full adaptation to suspensory locomotion in trees, like today's orangutan from Sumatra and Borneo, but quite different from the extant African great apes like gorillas, chimps, and bonobos.

This locomotor system also differed from that of the primitive hominoids *Kenyapithecus* and *Griphopithecus,* which were arboreal or semiterrestrial quadrupeds and bore a much more flexible vertebral column. In contrast, the lumbar vertebrae of *Dryopithecus* were short and placed within the thorax (not in the more dorsal position found in quadrupedal primates like *Proconsul* or *Kenyapithecus*). This means that *Dryopithecus* was already an orthograde primate; that is, it bore, like humans and the great apes, a rigid column, which is an adaptation to a vertical position of the body and a prerequisite for bipedal locomotion. Its shoulder blades, or scapulae, would have been on the back of the chest and not on the side, as in *Proconsul* and *Kenyapithecus*. As a suspensory ape, *Dryopithecus* had a humerus shaft that was straight and not curved, so the predominant movement of the arms would have been rotatory and above the head, rather than to the side and fore and aft.

Since *Dryopithecus* was, in fact, a rather archaic ape very close to the first hominoids that colonized Europe from Africa, we can expect that orthograde hominoids with a similar locomotor system were also present in Africa during the middle or late Miocene. If we assume this scenario, the evolution from quadrupedism to bipedism would have taken place not through a semiquadrupedal stage, like that of the African great apes (knuckle walking), but directly from the arboreal suspensory locomotion found in *Dryopithecus*. This would mean that the locomotion observed in today's gorillas and chimpanzees would be a secondary adaptation to life in the tropical forests of Africa.

CHAPTER 5

The Late Miocene:
The Beginning of the Crisis

ETWEEN 12 AND 11 MILLION YEARS AGO, A DRASTIC PULSE led to a new growth of the Antarctic Ice Sheet and a global sea-level fall of about 140 m. The oceans dropped about 90 m below the present sea level, and a number of land bridges again came into existence, thus enabling faunal exchanges between previously isolated terrestrial domains. Consequently, a corridor was reestablished between Asia and North America across what is now the Bering Strait. The main result of the reopening of this land bridge was the quick dispersal into Eurasia of the hipparionine horses of the genus *Hipparion* and their relatives.

The hipparionine horses arose in North America during the middle Miocene and differed significantly from *Anchitherium* and other early and middle Miocene equids in a number of respects. While the low-crowned, three-toed horses of the genus *Anchitherium* succeeded in the laurophyllous forests of Europe, in North America the extension of more open woodlands in the middle Miocene led to the appearance of new kinds of equids, characterized by their hypsodont dentition and slenderer limbs. This group, derived from *Merychippus,* developed high-crowned cheek-teeth as a response to the more sclerophyllous, harder vegetation. Moreover, the tooth enamel became folded in several ridges, which, in turn, filled with dental cement. With little variation, this kind of molar characterizes all modern horses, today's representatives of *Equus* included.

But changing environmental conditions in North America also induced significant changes in the equids' locomotor skeleton. The two persisting lateral toes in the hipparionine horses became more reduced than in *Anchitherium,* thus concentrating most of the body weight on the central toe. *Merychippus* and its close offshoot *Hipparion* were cursorial forms well adapted to the open woodlands that developed in North America. After the establishment of the Bering land bridge, the hipparionine horses quickly invaded the whole of Eurasia, from China to Iberia, their presence having been recorded in hundreds of fossiliferous localities. Existing data suggest that, after their entry into Eurasia, the hipparionine horses spread quickly across Europe, their fossils having been found from 11.1 million years ago in both in Vienna and Vallès-Penedès Basins. They probably colonized the more northern latitudes of Asia first and spread later to the south and east.

Hipparion primigenium, the first hipparionine species to enter Europe, was a relatively large form standing about 1.5 m at the withers (the stature of a Burchell's zebra) (figure 5.1). Its gracile build and axial skeleton suggest that *H. primigenium* was well adapted for leaping and springing, rather than for sustained running and high speed (Bernor and Armour-Chelu 1999). Some authors include this and other primitive hipparionine horses in the separate genus *Hippotherium.*

It is surprising that among the wide variety of large herbivores that inhabited North America at the beginning of the late Miocene (several camelids, bovids, and dromomerycids), only one equid species dispersed successfully into Eurasia when the Bering land bridge was reestablished. However, there is an explanation of *Hipparion*'s success over the several artiodactyl species present in Asia and North America at that time. As the paleontologist Christine Janis has shown, equids appear especially well suited to survive in poor habitats, where only low-energy food such as grass is available. In these environments, equids had an advantage over ruminants, being able to eat almost continuously. In contrast, ruminants cannot eat during rumination, which could have been a major disadvantage under some extreme circumstances. Although the first hipparionine horses that settled in Europe were basically browsers, the environmental conditions in the Bering area during

FIGURE 5.1 *Skeleton, musculature, and reconstructed life appearance of the equid* Hipparion primigenium

Exceptionally complete remains of this species, including several articulated skeletons, have been found at the site of Howenegg, Germany. Compared with a modern horse or zebra, *Hipparion primigenium* was a lightly built equid, and its body proportions were somewhat different: it had a slightly longer and more flexible back, and the forelimbs were shorter in proportion to the hind limbs. Such "deerlike" proportions point to an animal better adapted than modern horses to locomotion in forest and woodlands, which requires quick turns and sudden bursts of speed. In spite of these differences, *H. primigenium* would look much more horselike to a modern observer than the primitive anchitherine horses. Reconstructed shoulder height: 1.2 m

the cold pulse at 11 million years ago could have been particularly hard. At such a high latitude, the invaders probably had to face a rather poor diet based on grass, bush, and shrubs. Under these conditions, only an equid like *Hipparion* could have crossed and settled successfully.

Although a single taxon event, the dispersal of *Hipparion* dragged other immigrants from the open woodlands of central and western Asia into the laurophyllous forests of western Europe: the first European leporids of the genus *Alilepus,* the sivatherine giraffids of the genus *Decennatherium,* and the saber-toothed felids of the genus *Machairodus.* Among the lagomorphs, the leporids (the family including modern hares and rabbits) had had a long evolutionary history in North America since the Eocene. However, it was not until the early Vallesian that they settled in Europe, at that time still dominated by the pikas of the genus *Prolagus.* During most of the late Miocene and the Pliocene, the ochotonid *Prolagus* continued to be the most frequent lagomorph in any terrestrial European ecosystem, and it was not until the late Pliocene that the modern genera *Oryctolagus* and *Lepus* became common. Another immigrant, *Decennatherium,* was one of the first members of the sivatherines, a lineage of large, robust giraffids that differed from the slenderer paleotragines in the possession of not two but four ossicones, one pair over the orbits and the second, larger pair departing from the rear of the skull. The sivatherines became the dominant giraffids of the late Miocene terrestrial ecosystems and persisted in Africa until the early Pleistocene, coexisting with the first hominids. Like other primitive sivatherines, such as *Bramatherium* and *Hidaspitherium* from the Siwalik Hills, Pakistan, *Decennatherium* probably had an enlarged anterior pair of ossicones (or a unique fused anterior ossicone) and a less-prominent posterior pair. Its limb bones were longer and slenderer than those of the later members of this group.

Another early Vallesian newcomer was *Machairodus,* a large saber-toothed cat that gives the name to the subfamily of the machairodontines and coexisted with the last nimravids of the genus *Sansanosmilus* (figure 5.2). Apart from the large amphicyonids, all the other middle Miocene hypercarnivorous predators were relatively small, less than 100 kg, but members of the genus *Machairodus* were large saber-toothed cats that could attain 220 kg (about the size of a lion). Like that of the nimravids, the most characteristic feature of these predators was their long upper canines, which largely surpassed the size of the lower ones. Unlike those of modern cats, these canines were not conical but laterally compressed and flattened. Besides their long canines, these hypercarnivorous felids were endowed with bladelike cheek-teeth well adapted to slicing meat. The first machairodontine cats were recorded during the middle Miocene in the Greek-Iranian Province and persisted there until

FIGURE 5.2 *Reconstructed life appearance of the saber-toothed cat* Machairodus aphanistus For decades, this early species of *Machairodus* was known only on the basis of fragmentary remains. The discovery in the early 1990s of the carnivore trap site of Cerro Batallones, Spain, has provided nearly complete remains of several individuals, showing that this was a rather unspecialized, lion-size machairodontine. The body proportions were not unlike those of a living tiger, with long hind limbs well adapted to jumping. Reconstructed shoulder height: 1 m.

the early Vallesian (*Miomachairodus pseudailuroides* from Yeni-Eskihisar and Eçme-Akçacoy in Turkey). *Machairodus aphanistus* was the most common late Miocene species in Eurasia, ranging from the Iberian Peninsula to North America (where it was described as *Nimravides catacopis*). This species rapidly split into a number of offshoots, such as *M. romeri*, *M. copei*, and *M. alberdiae*. Several complete skeletons found at the Vallesian locality of Cerro Batallones in Madrid have revealed that *M. aphanistus* had rather tigerlike limb proportions, and thus particularly good jumping abilities. However, in other respects, *Machairodus* differed greatly from the modern large cats. For instance, the skull was straighter and more elongated, with narrow zygomatic arches and smaller eyes. Moreover, the neck was longer and with a strong musculature, as shown by the great development of the temporal crests. The forelimbs were very robust, with huge claws, and capable of considerable lateral rotation, enabling these predators to grasp and immobilize their prey, while the hind limbs were relatively long and well adapted to propel the animal's body in a sudden, short dash from ambush.

At first glance, the entry of *Machairodus*, a felid joining the large-predator guild previously occupied by *Sansanosmilus* alone, would appear to have had serious consequences for these nimravids, including their extinction by competition. However, this was not the case, and both *Machairodus* and *Sansanosmilus* coexisted during more than 1.5 million years without either one's replacing the other. A similar situation occurred in the Greek-Iranian Province, where the large nimravid *Barbourofelis* coexisted with *Miomachairodus,* as indicated by the early Vallesian beds of the Sinap Formation in Turkey. *Barbourofelis* was larger than *Sansanosmilus,* the size of a lion, and in the late Miocene attained a broad distribution, from North America to eastern Europe. But the dispersal of *Machairodus* into Europe coincided with the extinction of the two larger species of *Pseudaelurus, P. lorteti* and *P. quadridentatus,* which came to an end after millions of years of successful evolution through the early and middle Miocene (the more archaic *P. tournauensis* persisted until the early Turolian). In fact, newcomers like *Machairodus, Alilepus, Hipparion,* and *Decennatherium* joined the already highly diversified faunas that inhabited the middle Miocene subtropical forests of western Europe without a clear and immediate replacement of the potential competitors that were occupying their guilds there.

Despite its significant zoogeographic importance, the spread of the hipparionine horses and their cohort of Asian immigrants was a limited event that did not result in a significant change in the structure of the previously existing western European mammalian communities. This is even more surprising if we consider the climatic and oceanic events that took place at the beginning of the late Miocene. However, a similar situation has been reported from western Asia, where no significant mammalian turnover was associated with the first entry of *Hipparion* (Barry et al. 1985; Pilbeam et al. 1997).

THE EARLY VALLESIAN CLIMAX

The early Vallesian times in Europe are, therefore, characterized by the "peaceful" coexistence of a number of species that seem to have filled similar ecological guilds. This led to a sort of "climax" situation in the western European ecosystems, which reached levels of mammalian diversity unknown during any other late Cenozoic epoch. With more than sixty mammalian species, localities such as Can Ponsic and Can Llobateres 1, Spain, and Rudabanya, Hungary, are good examples of these early Vallesian "inflated" faunas. Despite the presence of new immigrants like *Hipparion, Decennatherium,* and *Machairodus,* the western European Vallesian ecosystems were composed of the same genera that populated the middle Miocene subtropical forests, retaining a similar community structure.

At the top of the large-browser guild were the proboscideans, represented by the large gomphotheres of the genus *Tetralophodon* and the deinotheres of the genus *Deinotherium.* Chalicotherid (*Chalicotherium*) and rhinocerotid perissodactyls still filled most of the large-browser guild. The most common rhino at that time was *Aceratherium*. This hornless aceratherine, close to the middle Miocene *Hoploaceratherium,* was one of the longest-lasting genera of the late Miocene, surviving until the Miocene–Pliocene boundary 5 million years ago. It was a medium-size rhino with long limbs and a still functional fifth metacarpal. The cheek-teeth were brachydont, indicating a browsing diet based on leaves and soft vegetables. Its limb proportions, close to those of today's tapirs, suggest a similar lifestyle. The males of *Aceratherium incisivum* bore a pair of strong incisors that enabled them to browse the dense vegetation of the early Vallesian laurophyllous woodlands (figure 5.3). These incisors were much smaller in the females, this being a clear case of sexual dimorphism among rhinoceroses.

Another group of perissodactyls to flourish at this time were the tapirs (*Tapirus priscus*), which reappeared in western Europe after their demise in the early Miocene. The anchitherine horses included larger, more-advanced species of *Anchitherium* that developed longer limbs and higher-crowned dentition. These anchitherine horses persisted in central Europe until the latest Vallesian, although in some regions, such as Spain, they disappeared shortly after the entry of the first hipparionine horses.

Among the smaller browsers, a peculiar group that settled in Europe were the hyraxes, which are found in a number of Vallesian localities, such as Can Llobateres in Spain; Melambes in Crete, Greece; and Eçme-Akçacoy, Turkey. Hyraxes are small ungulates that today inhabit the rocky and steppe environments of central and southern Africa (although their range extends up to Lebanon). Despite their small size and rabbitlike appearance, they are truly ungulates whose archaic morphology and anatomy resemble that of some condylarths. In fact, as noted in chapter 2, a number of authors regard the hyraxes as the "sister"-group of all living perissodactyls. They bear chisel-like incisors, which superficially resemble those of rodents, as well as a complete set of premolars (four) and molars (three) without diasteme. The molars are brachydont and selenolophodont, strongly resembling those of some archaic perissodactyls. Although modern hyraxes are hare-size (about 50 cm in length), the late Miocene European forms such as *Pliohyrax* reached considerable body dimensions, comparable to those of a modern tapir. The first hyraxes to settle in Europe during the early Vallesian certainly had an African origin. Therefore, a limited interchange with northern Africa still existed in the early Vallesian, although the possibility that hyraxes entered with the dryopithecids at the end of the middle Miocene cannot be excluded.

A wide array of suids of middle Miocene origin, such as *Listriodon, Propotamochoerus,* and *Parachleuastochoerus,* populated the undergrowth

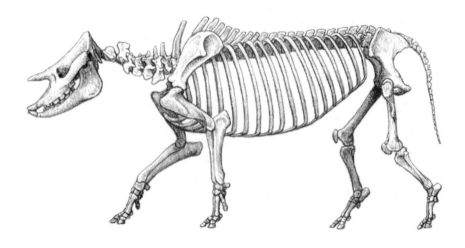

FIGURE 5.3 *Skeleton and reconstructed life appearance of the hornless rhinoceros* Aceratherium incisivum

Once more, the most complete skeletons of this species come from the site of Howenegg. The feet were preserved complete, including the vestigial fifth digit of the manus, which is lost in modern rhinos. The mandible displayed elongated incisors that give the species its name, but in spite of their length it seems likely that they were covered in life by the well-developed lips, as with modern Asiatic rhinos, which also may have quite long incisors. The cheek-teeth morphology suggests an essentially browsing habit, so we infer that the upper lip would have been long and vaguely hook-shaped, as in modern browsing rhino species. Reconstructed shoulder height: 1.3 m.

of the early Vallesian laurisilva. Also, moschoids like *Micromeryx* and *Hispanomeryx* and tragulids like *Dorcatherium* testify to the persistence of humid conditions in western and central Europe. A new member of the cervid family, *Amphiprox*, joined the middle Miocene *Euprox* in the early Vallesian ecosystems. *Amphiprox* was a small cervid of about 25 kg that resembled *Euprox* in bearing rather simple, two-forked antlers. Its teeth suggest that it was a browser with omnivorous tendencies. However, it had moderately long limbs with a long radius, indicating a good adaptation to open and light forests and, perhaps, to a more mountainous habitat.

Among the carnivores, the large amphicyonids of the genus *Amphicyon* persisted, represented by *A. major*. A second species, *A. castellanus*, was present in some localities in central Spain. Although similar in size to *A. major* (about 220 kg), *A. castellanus* was a hypercarnivore rather than a bone-cracker, as shown by the long, slicing blade of its lower molars. The hypercarnivorous and cursorial amphicyonids *Thaumastocyon dirus* and *Agnotherium antiquus* also persisted into the early Vallesian. *A. antiquus* is a poorly known species present in several localities from western and central Europe (Pedregueras, Eppelsheim, Rudabanya) and northern Africa (Bled Douarah). It was similar to the better-known middle Miocene *A. grivensis*, but smaller (160 kg). Finally, the small omnivorous amphicyonid *Pseudarctos bavaricus* also persisted into early Vallesian times.

Among the bears, the small omnivorous ursids of the genus *Ursavus* diversified into the species *U. brevirhinus* and *U. primaevus*, while the mesocarnivore *Hemicyon* included *H. goeriachensis* (of about 120 kg). But the most significant event among this group was the appearance of the first large ursids of the genus *Indarctos*. *Indarctos vireti*, of about 175 kg, was the first member of a lineage of large mesocarnivorous ursids that were characteristic forms of the late Miocene community of carnivores.

During the early Vallesian, the civetlike *Protictitherium*, the mongoose-like *Plioviverrops*, and the wolflike *Thalassictis* still represented the true hyaenids (figure 5.4). At this time, *Protictitherium* also colonized northern Africa, being present during the early Vallesian in Tunisia (*P. punicum*). The cursorial canidlike hyaenid guild was enriched with new forms close to *Thalassictis*, such as *Ictitherium* and *Hyaenotherium*. Both may have originated in the Greek-Iranian Province or elsewhere in Asia and spread into Europe. *Ictitherium* was, in fact, more omnivorous than *Thalassictis*, with less-sectorial carnassials, larger premolars, and less-reduced posterior molars. *Hyaenotherium* was close to *Thalassictis*, but larger (figure 5.5). The early Vallesian also records a high diversity of old viverrids (such as *Semigenetta ripolli* from Can Llobateres) and mustelids, which inherited the variety of forms present during the middle

FIGURE 5.4 *Two hyaenids of the species* Proticitherium crassum *attempt to scavenge from the* Hipparion *(*Hippotherium*) kill of a large saber-toothed cat,* Machairodus

Again from the site of Cerro Batallones come remarkably complete skeletons of the primitive hyaenid *Proticitherium crassum*, previously known from only fragmentary remains. The animal was about the size of a fox, with a locomotor apparatus closely resembling that of modern terrestrial civets. The dentition suggests a varied diet, including small vertebrates and a considerable amount of carrion, not unlike that of a modern jackal. This species persisted from the early Miocene into the Vallesian with little appreciable change. Reconstructed shoulder height of *Proticitherium*: 30 cm.

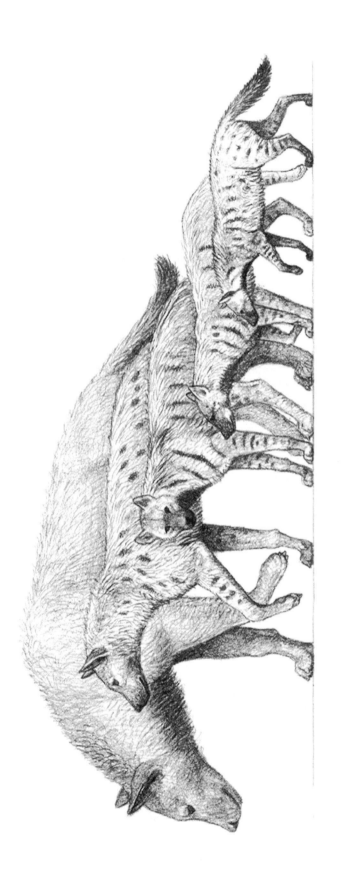

FIGURE 5-5 *Representative species of hyaenids from the late Miocene in Europe*

Left to right: Adcrocuta eximia, Hyaenotherium wongii, Ictitherium viverrinum, Protictitherium crassum, *and* Plioviverrops orbignyi. *Reconstructed shoulder height of* Adcrocuta: 80 cm.

Miocene: badgers (*Sabadellictis*), skunks (*Promephitis, Mesomephitis*), otters (*Sivaonyx, Limnonyx, Lutra*), and wolverines and glutton-related forms (*Trochictis, Circamustela, Marcetia, Plesiogulo*). Some of these glutton-related forms were relatively large for a mustelid, reaching 50 kg in the case of *Hadrictis* and *Eomellivora*.

Meanwhile, in the Greek-Iranian Province the hyena-like percrocutids persisted into the early Vallesian, represented by the genus *Dinocrocuta* (figure 5.6). They were larger than the middle Miocene *Percrocuta* and displayed clearer dental adaptations to bone-cracking. Like *Percrocuta* during the middle Miocene, *Dinocrocuta* coexisted during the early Vallesian with slender, cursorial true hyaenids such as *Thalassictis*. But the changing conditions at the beginning of the Vallesian had more severe effects on other groups in the Greek-Iranian Province. The elasmotherines and other eastern rhinoceroses became extinct, being replaced by grazing rhinoceroses of Asian and African origin. From the Asian side, members of the Central Asian aceratherine hornless genus *Chilotherium* became the most common rhinos in the eastern regions. They were a group of grazing animals that occupied different niches and radiated into a number of (sub)genera and species, such as *Subchilotherium* and *Acerorhinus*. Their legs were shorter than those of any other aceratherine, mimicking in some way those of the teleocerathines. The manus, however, was already tridactyl. Few of them were still clear browsers, like the brachydont *Acerorhinus,* while most of them were grass-eaters (although their diet certainly included a number of non-gramineous herbs). The shortening of the legs in this group can be explained on the basis of this grass-based diet. As aceratherines, they were hornless rhinos equipped with tusklike incisors, which they probably used in their fights. In contrast with that of today's grass-eating rhinos, the head maintained a horizontal position, so grazing was possible only after the shortening of the legs.

During the late Miocene, other advanced rhinos invaded this area from Africa and joined *Chilotherium* and other Greek-Iranian aceratherines. A clear example is *Ceratotherium,* the genus that includes the modern white rhino. The early representatives of *Ceratotherium* were only partly grass-eaters, but a trend is observed in this group throughout the Miocene and the Pliocene to develop more open-land adaptations, such as large body dimensions and slightly hypsodont cheek-teeth. Besides these grazing rhinos, other African immigrants to the Greek-Iranian Province, such as *Diceros pachygnatus,* a browsing rhino related to today's black rhino, point to the persistence of some densely vegetated areas in the late Miocene.

THE VALLESIAN HOMINOID RADIATION

Under these favorable conditions, the hominoids experienced a rapid diversification, being represented by a large variety of types. In western

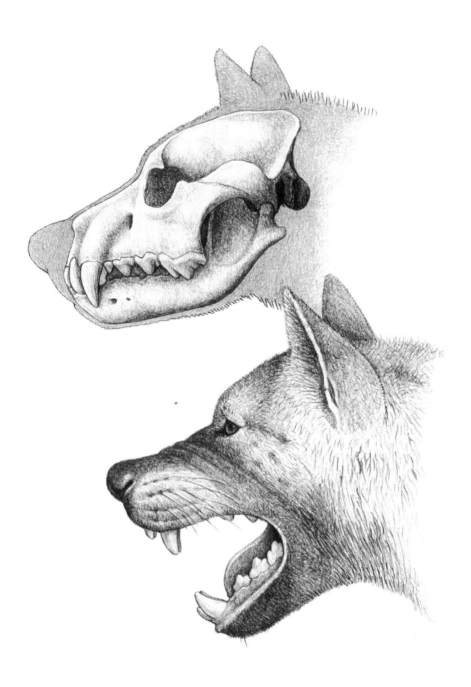

FIGURE 5.6 *Skull and reconstructed head of the giant percrocutid* Dinocrocuta gigantea

Members of the genus *Dinocrocuta* are known from fragmentary remains in Europe, but from the late Miocene in China complete skulls have been found, showing that although these percrocutids had paralleled true hyaenids to a remarkable degree, their skull morphology was rather unique. The cranium was very robust, with powerful zygomatic arches and a strongly domed forehead, and the teeth, especially the premolars, were proportionally enormous. Total length of skull: 40 cm.

and central Europe, *Dryopithecus* included at least two species: a larger
one, *D. fontani* (the original species described at Saint Gaudans), and a
smaller one, *D. laietanus,* of about 20 kg (described in the Vallès-Penedès
Basin and abundantly represented in several localities of this basin, the
CL-18000 skeleton from Can Llobateres 2 included) (figures 5.7 and 5.8).
According to some authors, a third species, *D. crusafonti,* was also present
in this basin (locality of Can Ponsic). *D. hungaricus,* from the rich Vallesian
site of Rudabanya, probably also belonged to this clade.

Coinciding with the spread of *Dryopithecus* and other hominoids during the early Vallesian, the pliopithecids became rarer in western Europe, although this group (*Anapithecus*) is well represented in Rudabanya, where it seems to have coexisted with *Dryopithecus* (Kordos,
personal communication, 1998). However, this seems to have been
more the exception than the rule, and this apparent exclusion between
the two kinds of primates suggests that they occupied different biotopes.
Probably, the folivorous primates like *Pliopithecus* remained as basic
dwellers of the lowland sclerophyllous forests, in which seasonality was
accentuated throughout the middle Miocene, while *Dryopithecus* populated the evergreen highland forests of western Europe, which probably
retained most of their humidity throughout the year, as do today's
broad-leaved laurophyllous forests in the Canary Islands. Since these
primates probably populated different altitudinal environments, the
abundance of one or the other kind in a locality would have depended
on the altitudinal boundary between the lowland sclerophyllous forests
and the highland laurophyllous forests. For instance, the abundance of
thin-enameled hominoids in localities such as Rudabanya, Can Ponsic,
and Can Llobateres 1 strongly suggests the extension to lower altitudes
of this evergreen forest during the late Aragonian and early Vallesian,
rather than the effect of purely taphonomic factors.

In the Greek-Iranian Province, *Dryopithecus* seems to be absent, replaced by a group of thick-enameled hominoids that were originally
described as *Ankarapithecus meteai*. As with *Dryopithecus,* little was known
until recently about *Ankarapithecus*—since the first discoveries in the
Sinap Formation, Turkey, during the 1950s—only that its dentition,
characterized by thick enamel, resembled in some ways that of the Asian
genus *Sivapithecus*. The description in 1985 of a fragmentary face with
the palate and part of the orbit showing some features close to the
orangutan clade supported the assignment of the Turkish hominoid to
Sivapithecus. But the discovery in 1996 of a rather well preserved skull
with its complete face led to a reassignment to the original genus, *Ankarapithecus*. *Ankarapithecus* was a small hominoid of about 23 to 29 kg
(that is, lighter than a bonobo). The orbits were large and more or less
circular, more similar to the African hominoids than to the narrow and
tall orbits of *Sivapithecus* and the orangutan. However, the interorbital

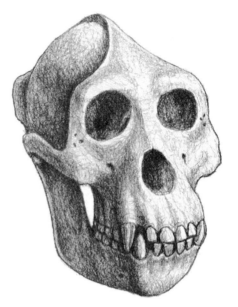

FIGURE 5.7 *Skull and reconstructed head of the hominoid* Dryopithecus laietanus

The best cranial remains of *Dryopithecus* were found at the site of Can Llobateres, Spain. The skull shows many characteristics that seem to be primitive for the Hominoidea, but there are also several features that suggest incipient evolution toward the modern orangutan.

breadth was rather narrow, as in the Asian hominoids. *Ankarapithecus* had a powerful masticatory apparatus, probably used for chewing hard food like seeds and other tough materials. This is in agreement with its robust, thick-enameled dentition, which was composed of big central incisors and relatively small canines. The presence of large central in-

FIGURE 5.8 *Reconstructed life appearance of* Dryopithecus laietanus

Also from Can Llobateres come the most complete skeletal remains of this species, providing for the first time an accurate picture of body proportions, posture, and locomotion in these early hominoids. The arms and, especially, the hands were remarkably long and showed early specialization toward suspensory habits. The locomotion of *Dryopithecus* in the trees would have resembled to a high degree that of today's orangutans.

cisors is a characteristic found also in some of today's hominoids like the orangutan and was probably related to food preparation before ingestion. Molars and premolars had poorly developed shearing crests and were large—a feature commonly known as megadonty. Megadonty increases the processing area of the teeth and was usually found in the early hominids of Africa, where the extension of a savanna-like environment forced a change in the diet toward consuming large quantities of plant foods. It is supposed that *Ankarapithecus* had to face rather similar conditions in the open-woodland environment of the Greek-Iranian Province.

Close to *Ankarapithecus,* but somewhat younger (about 9.6 million years old), was *Ouranopithecus* (also known as *Graecopithecus*) (figure 5.9). The first remains of this hominoid were described from the Greek locality of Pyrgos in 1972. Although, as with *Ankarapithecus,* little is known about the locomotor skeleton of *Ouranopithecus,* the affinities of this genus were in part clarified after the finding of a rather complete face—including the palate, nasals, and a complete orbit—at the locality of Xirochori. This discovery showed that *Ouranopithecus* (or *Graecopithecus*) was certainly bigger and more robust than *Dryopithecus* and *Ankarapithecus.* It was a rather large hominoid whose males could attain the size of a female gorilla. The face showed a robust aspect that superficially resembled that of the gorilla, with low and quadrangular orbits and a large interorbital region. A well-developed supraorbital torus bounded the orbits. The mandible and the maxillary bones were robust, which, as with *Ankarapithecus,* was related to the presence of large, megadont premolars and molars with thick enamel. The canines were relatively reduced with respect to such modern apes as the gorilla and chimpanzee. Although almost nothing is known about its locomotor habits, the microstructures left on the enamel reveal marks left by small silica grains during mastication. This is a feature of those primates that have to take their food not directly from trees but from the ground, where the seeds and bulbs collect a considerable quantity of dust and sandy particles. According to this analysis, *Ouranopithecus* was probably a ground dweller, unlike the western and central European dryopithecids.

All these features, but especially some characteristics of the dentition, led Louis de Bonis and Georges Koufos to propose *Ouranopithecus* as a feasible relative ("sister-group") of the first African hominids of the genus *Australopithecus.* The absence of any appropriate record of late Miocene African hominoids that could fit as such a plausible ancestor moved some authors to propose a kind of "back to Africa" scenario (in contrast to the familiar "out of Africa" scenario, which occurred during the Pleistocene). According to this new scenario, most of the genera that were going to characterize the Pliocene savannas of eastern Africa were already present in the Greek-Iranian Province during the

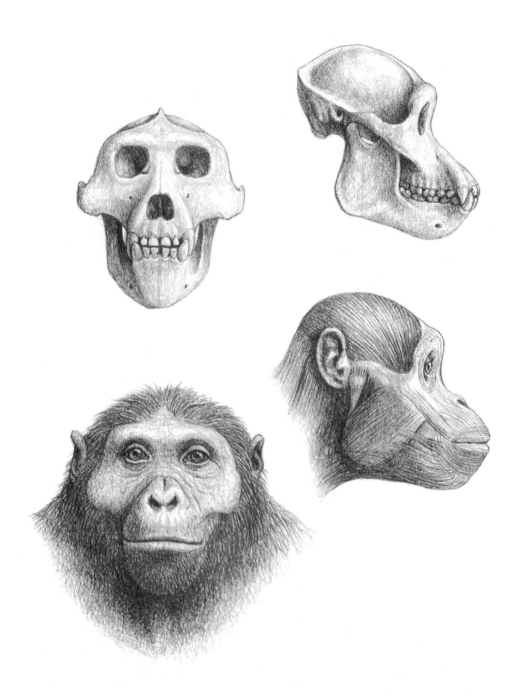

FIGURE 5.9 *Skull and reconstructed head of* Ouranopithecus (= Graecopithecus) ma-cedoniensis *in lateral and frontal views*

A remarkable partial skull of a male of this species was found in the locality of Xirochori, Greece, providing a more complete picture of the cranial anatomy of these animals. The Xirochori skull is remarkably robust, with a powerful masticatory apparatus somewhat reminiscent of that of Pliocene and Pleistocene hominids from Africa.

late Miocene. Therefore, the Greek-Iranian Province could have acted as the source that nourished the early Pliocene faunas of Africa, including, of course, the homin(o)ids. According to these authors, the open woodland that was present in that area during most of the Miocene may also be regarded as a feasible precedent of the early savanna (or proto-savanna, according to the term coined by Judith Harris [1993]). This open-woodland biome with most of its inhabitants could have extended to eastern Africa during the late Miocene or early Pliocene, establishing the basis for the later mammalian (and hominid) evolution on this continent.

However, a different interpretation tends to relate *Ouranopithecus* more to the Eurasian hominoids like *Ankarapithecus* than to the early African hominids. Although much more robust, *Ouranopithecus* shared a number of features with *Ankarapithecus*. In fact, the skull of Xirochori (upon which most of the phylogenetic analysis of this genus is based) belonged to a male, while the face of *Ankarapithecus* from the Sinap belonged to a female. Taking into account the usual sexual dimorphism present among modern apes, the distance between *Ouranopithecus* and *Ankarapithecus* may have been much closer than initially thought. So *Ouranopithecus* could have been part of the Eurasian clade that includes *Ankarapithecus* and *Sivapithecus* and that led to today's orangutan.

In any case, it is evident that after the settlement of generalized hominoids like *Griphopithecus* and *Dryopithecus* in Europe and western Asia during the middle Miocene, this group experienced its own evolutionary radiation, leading to a variety of genera that included arboreal quadrupeds, gracile suspensory forms, and robust, gorilla-like ground dwellers. In some ways, this evolutionary radiation paralleled the one that 5 million years later would produce in Africa present-day apes and the first bipedal hominids. The end of the story, however, was very different in Eurasia.

THE VALLESIAN CRISIS

After the high diversity levels attained during the early Vallesian, an abrupt decline in the Vallesian mammalian faunas took place at about 9.6 million years ago, in what is known as the Vallesian Crisis (Agustí and Moyà-Solà 1990). The Vallesian Crisis was first recognized in the Vallès-Penedès Basin of Spain and involved the sudden disappearance of most of the humid elements that had characterized the middle Miocene and early Vallesian faunas from western Europe. Among the large mammals, this crisis particularly affected several groups of perissodactyls, such as the rhinoceroses *Lartetotherium sansaniense* and *"Dicerorhinus" steinheimensis* and the tapirs (only the small tapirs of the little-known *Tapiriscus pannonicus* persisted until the early Turolian in central Europe). Partly compensating for these losses, *Stephanorhinus schleiermarcheri*, a large species of browsing rhino that persisted until the Miocene–Pliocene boundary and is considered by some authors to be the first member of the lineage leading to the European Pleistocene rhinocerotids, came on the scene just before the onset of the Vallesian Crisis. *S. schleiermarcheri* bore a pair of massive horns and was the largest rhino of its time. Among the artiodactyls, the huge diversity attained by the suids during the early Vallesian suddenly dropped, and several characteristic groups vanished, such as the browsers *Listriodon* and *Schizochoerus* (which made a short-lived incursion into western Europe at the beginning of the late Vallesian), as well as the tetraconodontines *Conohyus* and *Parachleuastochoerus*. In contrast, the "modern" suinae such as *Propotamochoerus* persisted and even enlarged their diversity with new eastern immigrants like *Microstonyx* (figure 5.10). *Microstonyx* was a giant pig (about 300 kg) with a skull more than 0.5 m long. However, as a member of the modern suinae, it was close in many other respects to today's pigs. The Vallesian Crisis also involved the final decline of the middle Miocene forest community of cervoids (the cervid *Amphiprox* and the moschid *Hispanomeryx*) and the spread of the boselaphine bovids like *Tragoportax*, which replaced their semiaquatic relatives of the genus *Protragocerus*. *Tragoportax* was a medium-size bovid of about 80 kg with relatively long limbs, which suggests that it was a fast runner and a good jumper that lived in the open woodland (figure 5.11). It possessed a short-faced skull with long neurocranium and large backward-curving horns. The horn cores of the females and young were smaller and slenderer than those of the males (in fact, the young individuals of *T. gaudryi* were originally described as a different genus, *Graecoryx*). The teeth were high crowned and with cement, resembling those of today's *Boselaphus*.

Among the rodents, the Vallesian Crisis involved the disappearance of most of the cricetids and glirids of early or middle Miocene origin (*Megacricetodon, Eumyarion, Bransatoglis, Myoglis, Paraglirulus, Eomuscardinus*), flying squirrels (*Albanensia, Miopetaurista*), and beavers (*Chalicomys, Euroxenomys*). Other less-diversified small-mammal groups, such as the lagomorphs and insectivores, remained almost unaffected by this crisis. In western and central Europe, this event coincided with the first dispersal of the murid rodents, the family that includes today's mice and rats. After their entry into Europe, members of this group became the dominant rodents in the late Miocene communities and diversified into several genera: *Progonomys, Occitanomys, Huerzelerimys,* and *Parapodemus*.

Another significant event among the small-mammal faunas concerned the local evolution of the *Cricetulodon* hamsters, which developed a peculiar dental design, the so-called sigmodont pattern. *Cricetulodon* was the first recognizable member of the lineage leading to the hamsters that we enjoy today as pets. They entered Europe during the early Val-

FIGURE 5.10 *Skull, musculature of the head, and reconstructed life appearance of the giant suid* Microstonyx major

Fine skulls and isolated postcranials of this enormous pig have been discovered in many European sites, including Pikermi, Greece, and Piera, Spain. These discoveries show that, in spite of its huge size, this animal was closely similar to modern members of the suid family, such as the wild boar. The area for insertion of the muscles that retract the nose shows that there must have been a strong, highly mobile muzzle disk like that of modern pigs, which would have been used for digging roots and tubers. Total length of skull: 60 cm. Reconstructed shoulder height: 1.1 m.

FIGURE 5.11 *Skull, musculature, and reconstructed head of the boselaphine bovid* Tragopor-tax gaudryi

Excellent fossils of this species were recovered at the site of Piera, Spain, including complete skulls and mandibles. The Piera sample has shown that these antelopes were sexually dimorphous, with the males, like the one depicted in this restoration, displaying longer and heavier horns than the females. Total length of skull: 26 cm.

lesian and became the most common rodent species in several localities of western and eastern Europe. During the late Vallesian, some species of this group, still bearing low-crowned, bunodont cheek-teeth, showed a quick trend toward developing sigmodont-semihypsodont molars (these advanced species are usually included in a separate genus, *Rotundomys*). Rodents displaying the "sigmodont pattern" present a dental design based on a unique sinusoidal ridge and a flat wear surface. This design is present in several extant rodents and is usually associated with a grazing diet (for instance, in the Holarctic arvicolids and the Nearctic "cotton rats"). This evolutionary innovation involved not a big change in the basic cricetid dental morphology (except for the development of a flat wear surface), but a profound functional change in the chewing process. From the original low-crowned, bunodont design, in which the transverse movements were dominant, these rodents developed a more complex chewing process, with a clear anterior–posterior component during mastication.

The evolutionary change from *Cricetulodon* to *Rotundomys* was a highly significant event that probably indicates the spread of a more sclero-phyllous vegetation or even grasses. In fact, the appearance of sigmodont patterns in several cricetid lineages over a wide area characterized the transition from the early to the late Vallesian. This occurred in *Microtocricetus* (from several early and late Vallesian localities in central Europe; Franzen and Storch 1999) and *Ischymomys* (from eastern Europe and western Asia; Nesin and Topachevski 1999).

Another group severely affected by the Vallesian Crisis was the large carnivores of the families Nimravidae and Amphicyonidae. Among the amphicyonids, all the genera and species still existing in the early Vallesian disappeared: *Pseudarctos bavaricus*, *Amphicyon major*, and *Thaumastocyon dirus*. Only some lesser-known *Amphicyon* representatives persisted into the late Vallesian and early Turolian in some parts of central Europe: *A. gutmanni* (246 kg) from Germany and Austria (Mannersdorf and Kohfidisch) and *A. pannonicus* (198 kg) from Hungary (Pecs). All these relics seem closely related to the early Vallesian *A. major*, although *A. gutmanni* displayed a more-robust and high-crowned carnassial. Like *A. major*, they were probably active hunters with bone-crushing habits. The presence of other carnivores that were either more catholic in their dietary requirements (for example, Ursinae) or more cursorially adapted (Hyaenidae) may have played a role in the extinction of these amphicyonids. However, most of the cursorial hyaenids also vanished.

Among the ursids, the Vallesian Crisis had an ambivalent effect. While the slender cursorial forms of early Miocene origin like *Hemicyon* vanished, the robust ursids of "modern" appearance persisted, represented by larger species. This was the case with *Indarctos*, represented by the species *I. vireti* and *I. arctoides*, as well as *Ursavus*, represented by

U. depereti, the largest species of the genus, in a number of late Vallesian localities, such as Soblay and Melchingen.

The mustelids were also severely affected by the Vallesian Crisis, suffering a significant decrease with respect to the once highly diversified Vallesian fauna.

At the same time, a number of eastern immigrants appeared for the first time, such as the large hyaenids of the genus *Adcrocuta* and *Hyaenictis* (figure 5.5). These genera represented two opposite trends in the evolution of late Miocene hyaenids. *Hyaenictis* was a cursorial meat- and bone-eater that prolonged the trend initiated by *Thalassictis* toward increasing cursoriality. *Hyaenictis* also showed a trend toward reducing the bone-crushing portion of the dentition, developing and extending at the same time the sectorial part, so that the posterior molars were reduced or lost. At the other end of the scale, *Adcrocuta,* at about 70 kg, was the first representative of the modern bone-cracker hyaenids leading to today's *Crocuta* and *Parahyaena.* It was characterized by advanced adaptation to bone-crushing, with enlarged bone-cracking premolars. *Adcrocuta* had short, stocky limbs, indicating that it was not a cursorial form. Like *Hyaenictis,* it was probably of Asian origin (both appear in the Spanish site of Viladecavalls together with eastern immigrants such as *Schizochoerus*).

Among the large predators, the nimravids finally came to an end during the Vallesian Crisis, after representing the "large-cat" guild during millions of years. The entry of *Paramachairodus,* a new genus of machairodontine cats, compensated for the extinction of the nimravids. *Paramachairodus ogygya,* the oldest species of this genus, was smaller and slenderer than *Machairodus aphanistus* (about 44 kg), retaining the archaic anatomy inherited from its close relative *Pseudaelurus quadridentatus.* With their robust forelimbs and gracile hind limbs, the members of *P. ogygya* were probably able to climb trees carrying large prey, as do modern leopards. This species's long-muzzled skull was also superficially leopardlike, although, like those of a true machairodont, the upper canines were characteristically long and laterally flattened. One particularity of this species was that the lower canines were also relatively long, an unusual characteristic even for a machairodont. In this respect, it resembled today's clouded leopard (*Neofelis nebulosa*). *P. ogygya* was the first member of the lineage that led to *Smilodon,* the famous machairodontine cat represented by hundreds of skeletons at the late Pleistocene site of Rancho La Brea in California and immortalized by Charles R. Knight in the murals of the American Museum of Natural History in New York.

Last but not least, the Vallesian Crisis led to an abrupt end of the hominoid experiment in Europe. Hominoids like *Dryopithecus, Ankar-*

apithecus, and *Ouranopithecus* disappeared entirely from the fossil record, and only *Oreopithecus* in the island refuge of Tuscany and *Sivapithecus* in southwestern Asia survived this extinction event. *Dryopithecus* was still found in some early late Vallesian localities, dated at about 9.6 million years ago (Can Llobateres 2, Viladecavalls), but disappeared from the fossil record shortly thereafter. In the Greek-Iranian Province, the robust *Ouranopithecus* also disappeared at the beginning of the late Vallesian. The Turkish *Ankarapithecus* did not reach the end of the late Vallesian. The extinction of these robust hominoids in this province coincided with the spread of the colobine monkeys of the genus *Mesopithecus* (figure 5.12). This was not the case in western Europe, where the extinction of the slender dryopithecines did not coincide with its replacement by any other kind of primate species. Only the persistence until the latest Vallesian of the advanced folivorous pliopithecids of the

FIGURE 5.12 *Reconstructed life appearance of the colobine monkey* Mesopithecus pentelecicum

The sample of *Mesopithecus* fossils from Pikermi includes well-preserved skulls and postcranials, allowing us to reconstruct the animals' life appearance with confidence. The body proportions of *M. pentelecicum* closely resembled those of the modern Hanuman langur of India, which is the most terrestrial among living colobines. Several other features of the skeleton also betray adaptations to locomotion on the ground, which would obviously have been advantageous to a monkey living in a more or less open, seasonally dry woodland habitat. Reconstructed shoulder height: 35 cm.

genus *Anapithecus* can be cited in this area, some hundred thousands of years after the last *Dryopithecus*. In China, the pliopithecids survived even longer, until the latest Miocene, being represented by a large form (*Laccopithecus*).

CAUSES OF THE VALLESIAN CRISIS

What could have caused the set of extinctions and deep faunal restructurings that took place 9.6 million years ago during the Vallesian Crisis? Some evidence, such as the extinction of several forest forms and the development of sigmodont teeth by some groups of rodents, would support the replacement at that time of the laurophyllous forests by grasslands. However, Cerling and co-workers (1997) have dated the spread of grasses over large stretches of Eurasia between 8.3 and 7 million years ago, and the geochemical analysis carried out on teeth and soil nodules older than that do not detect any signal of such an environmental change. Nevertheless, the fact that this major ecological restructuring of the western European mammalian assemblages especially affected those taxa with tropical-forest affinities and the latitudinal character of these extinctions (a number of forest-adapted taxa survived until the early Turolian in central Europe) strongly suggest a climatic cause. Additional evidence comes from the late Miocene oceanic evolution, which was a continuation of the processes started at the middle to late Miocene transition. Enhancement of the latitudinal thermal gradient resulted in the generation of new erosive oceanic surfaces (NH5) by the intensification of deep circulation. Further coolings resulted in new $\delta^{18}O$ positive shifts (that is, Mi6 and Mi7). Changes of benthic and planktonic assemblages also indicate colder climatic conditions and increasing isolation between low and middle latitudes. All these oceanic changes were nearly synchronous with some significant changes in low-latitude European terrestrial domains. In particular, there is a noticeable synchronism between the Mi7 isotopic shift at 9.6 to 9.3 million years ago and the age of 9.6 million years obtained for the Vallesian Crisis in the Vallès-Penedès stratigraphic sections. The Vallesian Crisis was also close to the beginning of the NH5 hiatus, one of the most important sets of deep oceanic discontinuities recognized in the late Miocene (Keller and Barron 1983), which is dated between 9.5 and 9.0 million years ago. NH5 has been also related to a period of cooling and major restructuring of the deep oceanic circulation and to the growth of ice sheets in western Antarctica (Keller and Barron 1983).

But how did these changes affect the composition of terrestrial vegetation? As we have seen, a change to more open environments did not start until 8 million years ago, the extension of grasslands taking place almost 2 million years after the Vallesian Crisis. However, we know that the laurophyllous subtropical woodland prevailing until the beginning of the late Miocene in Europe was profoundly affected in some way. The answer arrived in 1994, when for the first time a well-calibrated late Vallesian flora was discovered in a section close to the city of Terrassa, in the Vallès-Penedès Basin. This flora has been dated by paleomagnetism at somewhat more than 9 million years old and, therefore, records the kind of vegetation that was dominant just after the Vallesian Crisis.

This was the first time that a flora of late Vallesian age was discovered in the type-area of the Vallesian, since most of the previously known floras in the basin were of middle Miocene or early Vallesian age. The pre-Vallesian floras in the Vallès-Penedès Basin indicated the persistence of humid subtropical conditions, with an abundance of the early and middle Miocene laurophyllous elements. Indeed, the locality of Can Llobateres 1 yielded the remains of some of these subtropical elements, such as *Sabal* and fig trees. However, the analysis of the Terrassa flora (Sanz de Siria 1997) showed a surprising result: the vegetation of the late Vallesian was quite different from that of the middle Miocene and was closer, in fact, to that of the early Pliocene. Close to 45% of this flora was composed of deciduous trees, such as maples, alders, oaks, hickories, walnuts, elms, and limes—that is, the trees that now dominate the temperate forests of the middle latitudes. In contrast, the warm subtropical vegetation—represented by *Myrsinie, Sapindus, Sapotacites,* and so on—decreased to 8%. Supporting the argument that this change was not related to the extension of grasslands was the persistence of a "hard core" of subtropical elements, represented by the 33% of evergreen trees (*Laurophyllum,* laurel, buckthorn, and others) that persisted in Europe until the early Pliocene. The persistence of these subtropical elements, able to endure a certain level of seasonality, confirms the continuation in Europe during the late Miocene of the levels of humidity that had prevailed during most of the Miocene. However, the presence at Terrassa of up to 15% of Mediterranean or pre-Mediterranean taxa (such as the evergreen oaks *Quercus* cf. *ilex* and *Q. precursor*) indicates the existence at that time of a dry season (summer drought).

Therefore, the flora from Terrassa confirm that the profound faunal change at 9.6 million years ago was not the consequence of the replacement of the subtropical Miocene forest by grasslands, but the substitution of one kind of woodland by another. The intensification of the thermal gradients between the middle and the low latitudes, probably enhanced by the Himalayan and Tibetan uplifts, led to an abrupt change in the previously existing evergreen subtropical woodlands of western Europe, which were replaced by a more seasonally adapted, deciduous woodland. Rather than the moderate cooling associated with the Mi7 isotopic shift, it was this change in the structure of the vegetation that determined the set of extinctions that took place during the Vallesian Crisis.

Most members of this fauna, including *Dryopithecus* and other European hominoids, were mainly frugivorous, with a diet based on fruits and the soft vegetables common in the evergreen laurophyllous woodlands of the Miocene. Although the decreasing temperature and increasing latitudinal gradient had little direct effect on the Vallesian mammals, the replacement of most of the evergreen laurophyllous trees by deciduous trees, well adapted to the new conditions of seasonality with colder winters and drier summers, had much more dramatic effects. A number of pigs, rodents, and primates had to subsist during several months without fruits, a basic and highly nutritional component of their diet. This factor, not the extension of grasslands or the shift of temperature, was probably the direct agent causing the abrupt drop of the rich early Vallesian faunas.

In turn, the sudden disappearance of most of the medium-size herbivores that had settled in Europe over millions of years probably led to a critical situation for the old predators of middle Miocene origin, such as the nimravids and the amphicyonids, which until the early Vallesian has successfully endured competition from the machairodont cats and the large ursids. These carnivores were also indirect victims of the vegetational change that took place 9.6 million years ago.

THE TUROLIAN

Crusafont (1950) defined the Turolian mammal stage based on the faunal composition of a number of European late Miocene localities that differed from middle Miocene and Vallesian ones in the abundance and variety of large herbivores, particularly hipparions and several bovid species. As happened with the middle Miocene *Hispanotherium* faunas from Spain, the Turolian brought about a significant uniformity between the faunas from western Europe and those from the Greek-Iranian Province, both domains sharing several species and ecotypes and maintaining a similar ecosystem structure (although the western European faunas were always less diversified than those from the east). Because of the dominance of the large herbivores (hipparions, antelopes, rhinoceroses, and proboscideans), the typical Turolian associations have usually been regarded as the first expression of a "savanna" biome in the Miocene, although this assignment has proved to be an oversimplification. Most probably, the Turolian marks the maximum extension to the west of the conditions of the Greek-Iranian Province, after the demise of forest-adapted forms during the Vallesian Crisis. The most classic and diversified Turolian sites are those of Pikermi and Samos, Greece, and Concud and Los Mansuetos in the Teruel Basin, Spain (hence the name Turolian). However, several other Turolian localities have been also recognized, from Iran (Maraghe) to central and western Europe (for instance, Mont Leberon, France, and Dörn-Durkheim, Germany).

The climax of the Turolian faunas in Europe also coincided with a significant environmental change that affected the composition of the vegetation. The geochemical analysis carried out in paleosoil and dental carbonates between 8 and 7 million years old indicates a change from a dominant C3 to a dominant C4 vegetation. Most of the trees and bushes that are common in forests and woodlands use a photosynthetic pathway known as C3. However, grasses and other plants characteristic of grasslands use a different, more efficient photosynthetic pathway, well adapted to the harder physiological conditions of arid environments. These are known as C4 plants. A number of analyses conducted on a worldwide scale have shown that between 8 and 7 million years ago there was a significant shift in the $\delta^{13}C$ isotopic composition of paleosoil and dental carbonates; this shift has been interpreted as an indication of the replacement of C3 plants by C4 plants and the spread of prairies and grasslands over Eurasia and Africa (Quade et al. 1989; Morgan et al. 1994; Cerling et al. 1997). The extension of the open-woodland faunas of the Greek-Iranian Province toward the west can be interpreted in the context of this environmental shift toward drier conditions and grassland-dominated biomes.

In the region of the Siwaliks, Pakistan, to the south of the Himalayas, the changes of 9.6 million years ago can scarcely be compared with the dramatic effects of the Vallesian Crisis in western Europe. Only a change in the relative proportion of tragulids (from 45% to 10%) and bovids (from 45% to 80%; Barry et al. 1991) is observed at this moment. A faunal association including archaic carnivores and aquatic rhinoceroses (*Brachypotherium*), proboscideans (*Deinotherium*), dormice, shrews, and hominoids with climbing locomotor adaptions (*Sivapithecus*) persisted until 8.3 million years ago in this area. This delay of more than 1 million years between western Europe and the Siwaliks in the onset of the late Miocene faunal turnovers is probably explained on the basis of the settling of monsoon atmospheric dynamics in the latter region, which could have maintained the forested subtropical conditions there until 8.3 million years ago. But between 8.3 and 7.8 million years ago, a major ecological change associated with a set of extinctions similar to those of the Vallesian Crisis took place in the Siwaliks (Pilbeam et al. 1996). This faunal turnover led to the final disappearance of several cricetid, bovid, and tragulid species, while *Sivapithecus* was replaced by colobine monkeys. For the first time, the Siwaliks fauna became similar to those of western Eurasia.

In western and central Europe, the persistence of some Vallesian survivors of humid character characterized the early Turolian at about 8.7 million years ago. These survivors included the deinotheres (which in the case of *Deinotherium gigantissimum* attained enormous dimensions, more than 4 m at the withers), chalicotheres (*Chalicotherium*), moschoids (*Micromeryx*), and tragulids (*Dorcatherium*). Their survival indicates the persistence of rather humid and forested conditions. They finally disappeared during the course of Turolian times, although some relatives of these Vallesian relicts survived outside Europe (for instance, deinotheres and chalicotheres persisted in Africa until the early Pleistocene, while moschids and tragulids still dwell in the dense forests of parts of Africa and eastern Asia).

But these early Turolian animals excepted, all the mammalian associations of this time indicate the extension of open woodlands and grasslands. There was an extraordinary diversification of bovids, particularly among the boselaphines *Miotragocerus* and *Tragoportax,* which included several species in the Greek-Iranian Province (figure 5.13). Also during the Turolian, the gazelles became widespread in the Mediterranean-Peritethyan region (from northern Africa and Spain to Turkey and the Ukraine), joining the boselaphines as the most common and most diversified late Miocene bovids. Other than boselaphines and gazelles, an extraordinary variety of spiral-horned antelopes, such as *Prostrepsiceros, Ouzoceros, Nisidorcas, Samotragus, Oioceros,* and *Hispanodorcas,* were also present. All of them were small gazelle-like antelopes weighing between 10 and 15 kg, differing mainly in the shape and position of the horn cores. *Prostrepsiceros* had rather long, divergent horns and relatively long limbs. The teeth were mesodont, indicating a diet based on hard leaves and herbs. It was a fast runner and a good jumper, inhabiting the dry bush shrubs and open woodlands. *Ouzoceros* was similar, but with less divergent and more upright horns. *Nisidorcas* was also a small antelope, but with short legs and less spiraled horns, that probably inhabited open country with bush shrub. In several respects, it resembled today's gazelles. The morphology of the skulls of *Samotragus* and *Oioceros* also resembled that of the gazelles, but the teeth of these genera were less specialized, probably indicating a diet based more on leaves than grass. They had long legs adapted to running over open ground.

Another group of bovids attaining a high diversity during the Turolian was the peculiar tribe of the Ovibovini. These animals, today represented by the musk ox (*Ovibos moschatus*) and the takin (*Budorcas taxicolor*), have been characterized throughout their evolutionary history by the particular specializations of their horn cores and basioccipital/atlas articulations, which include pronounced bone thickening in these regions. During the Turolian, the ovibovines in the Greek-Iranian Province included several genera, such as *Urmiatherium, Parurmiatherium,*

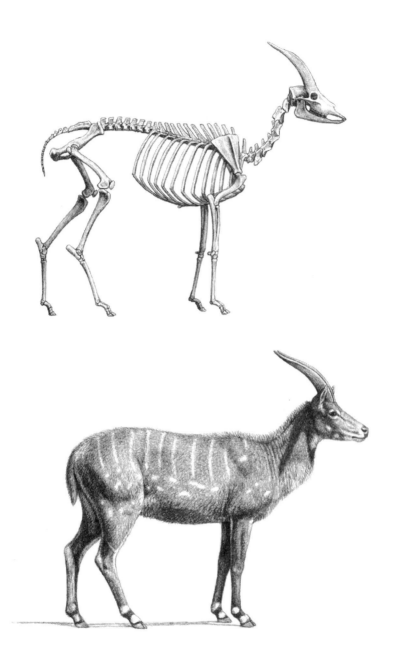

FIGURE 5.13 *Skeleton and reconstructed life appearance of the boselaphine* Miotragocerus panonniae

The best remains of this bovid, including articulated skeletons, come again from the site of Howenegg. Although the closest living relative of *Miotragocerus panonniae* is the Indian nilgai, the body proportions of the fossil species are more similar to those of such semiaquatic antelopes as the sitatunga and the kob, with a high rump, low shoulders, and splaying digits well adapted for locomotion on wet ground. It is probable that *M. panonniae* preferred to live near the water, where it could retreat for safety if threatened by predators. Reconstructed shoulder height: 1 m.

Criotherium, Budorcas, Plesiaddax, and *Mesembriacerus.* One of the most bizarre was *Plesiaddax,* a large bovid of about 150 kg with long limbs and body proportions similar to those of the Alcelaphini (the tribe that includes the African gnus, hartebeests, and impalas). The teeth were hypsodont and with cement, suggesting a diet based on tough plants that probably included leaves, herbs, and grass. The horns of *Plesiaddax* were short and departed laterally from the rear of the skull. Another Turolian Ovibovini, *Mesembriacerus,* was a medium-size bovid of about 25 kg, with long and slender horns strongly directed backward. It probably dwelt in open woodlands or more open habitats. Its dentition indicates a diet based on soft plants. It fits as a possible ancestor for the Pleistocene genera *Praeovibos* and *Ovibos.* Last but not least, the first goats (subfamily Caprinae, tribe Caprini or Hippotragini) appeared during the Turolian: *Palaeoryx, Protoryx,* and *Pachytragus* (plate 11). They probably evolved from a *Tethytragus*-like form during Vallesian times (in fact, *Tethytragus* fits as a possible ancestor for *Protoryx*).

The giraffids also attained an extraordinary diversity during the Turolian, especially in the eastern regions. The two-horned, slender paleotragines included several species: *Palaeotragus moldavicus, P. coelophrys,* and *P. roueni,* among others. Similar in morphology to *Palaeotragus,* but much larger and with more-hypsodont teeth, was *Samotherium.* The four-horned sivatherines also tended to evolve into large and robust forms, such as *Birgerbohlinia* in the west and *Helladotherium* in the east, which replaced the slender *Decennatherium* from the Vallesian (figure 5.14). Like *Decennatherium, Birgerbohlinia* bore a posterior pair of large ossicones and an anterior pair of smaller ones. *Helladotherium* was a large sivathere, with limb proportions similar to those of *Samotherium* but more robust individual bones. In later sivatheres, the metapodials became notably short. An opposite trend toward developing long, slender legs occurred among other coeval nonsivatherine giraffids such as *Bohlinia* (figures 5.15 and 5.16). This genus probably lies close to the first members of the lineage leading to today's *Giraffa.*

FIGURE 5.14 *Skeleton, musculature, and reconstructed life appearance of the sivatherine giraffid* Birgerbohlinia schaubi

While the postcranial skeleton of *Birgerbohlinia schaubi* has been known for decades, thanks to the finds at the site of Piera, the shape and position of the cranial appendages, or ossicones, was revealed only after the discovery in the late 1980s of an associated set of ossicones and cranial fragments at another Spanish site, Crevillente. Like other sivatherines, *Birgerbohlinia* was a robust animal, more similar in overall build to a large bovid such as an eland than to a modern giraffe or even an okapi. But the four-horned head would certainly make the appearance of these animals unmistakable to a modern observer. Reconstructed shoulder height: 1.8 m.

FIGURE 5.15 *Comparison between the heads of* Bohlinia attica (*bottom*) *and the modern giraffe,* Giraffa camelopardalis (*top*)

Although no complete skulls of *Bohlinia attica* are known, the available fragments, which include a partial skull with ossicones from Samos, Greece, provide a reasonable picture of the animal's head. Compared with a modern giraffe, *B. attica* had a longer, lower head, thus resembling the primitive paleotragine giraffids. The two ossicones placed over the orbits were somewhat more pointed than those of the giraffe, and the protuberance displayed in front of and between the ossicones in the extant species was absent in the fossil.

In contrast with the enormous diversity attained by bovids and gir-affids during the Turolian, diversity among cervids remained at low lev-els, represented by few genera such as *Lucentia, Turiacemas,* and *Proca-preolus* (plate 11). In general, Turolian cervids looked not much different from the Vallesian ones like *Euprox* or *Amphiprox.* The antlers continued to be rather simple, retaining the two-forked shape of their ancestors, but the pedicle was shorter while the shaft was longer, with increasing asymmetry between the forks. At the same time, the cervids tended to develop higher, more-hypsodont molars.

As with the bovids, the hipparionine horses attained their highest diversity in the middle Turolian, each locality having at least two or three species. In many cases, these sympatric species differed significantly in their size, each one occupying its own browsing and grazing level. The large, forest-dweller hipparions of the (sub)genus *Hippotherium* per-sisted, but were accompanied by the smaller and slenderer *Hipparion (sensu stricto).* This latter genus reached a high diversity during the Turolian, with several vicarious species ranging from Spain to Iran: *H. melendezi, H. gettyi, H. prostylum, H. fietrichi, H. capmbelli, H. concu-dense, H. antelopinum,* and others. All were medium-size, moderately high-crowned, slender hipparionines with elongated metapodials. They were probably active grazers adapted to somewhat open country. Pe-culiarly, some species of the (sub)genus *Cremohipparion* developed dwarf dimensions. *C. periafricanum,* from eastern Europe and Spain, was a small species less than 1 m at the withers. Like most of the Turolian hipparions, *Cremohipparion* was mainly a grazer (figures 5.17–5.19).

Among the rhinoceroses, the Turolian faunas consisted of the main groups that survived the Vallesian Crisis—that is, the hornless acerath-erines and the horned rhinocerotines. Among the former, *Aceratherium incisivum* persisted until the latest Miocene. *Alicornops* also persisted in Spain, represented by the species *A. alfambrensis.* In the Greek-Iranian Province, *Chilotherium* and other related eastern aceratherine rhinos like *Acerorhinus* showed increasing hypsodonty and specialization in grazing. The trend toward shortening the legs (by shortening the metapodials), as a way of bringing the muzzle nearer to the ground and facilitating grazing, persisted in this group (Heissig 1999b). Among the horned rhinocerotines, the group of "modern" rhinos that started with *Stephan-orhinus schleiermarcheri* persisted in the Greek-Iranian Province with the species *S. pikermiensis.* The "African" horned rhinos of the genera *Diceros* and *Ceratotherium* also persisted in this last region, represented by the species *D. primaevus* and *C. neumayri* (figure 5.20). The degree of hyp-sodonty of these fossil species suggests dietary preferences similar to those of their modern relatives, *D. primaevus* being basically a browser (like the black rhino) and *C. neumayri* being basically a grazer (like the white rhino).

FIGURE 5.16 *A water-hole scene from the Turolian site of Pikermi, Greece, with the giraffid* Bohlinia attica *and the hipparionine horse* Cremohipparion mediterraneum

In spite of differences in the shape of the head, the body proportions of *Bohlinia attica* were remarkably similar to those of modern giraffes, and they were just as tall. At Pikermi, they would have browsed in the more or less open woodland that dominated the landscape, venturing regularly to the more exposed areas around water holes in order to drink. Grazers and mixed feeders like the hipparions would favor edaphic grasslands around fluctuating water bodies, with gatherings of different ungulate species becoming a common sight around drinking places. Reconstructed total height (to top of skull) of *Bohlinia*: 5.3 m.

Another peculiar element among the Turolian Greek-Iranian faunas were the aardvarks, represented at the site of Samos by the species *Orycteropus gaudryi*. Present-day *O. afer* are archaic nocturnal ungulates that live in the open woodlands and grasslands of Africa. They have a superficial resemblance to pigs (hence the name "earth pig") because of their tubular snout, which ends in a blunt muzzle. Their diet is based on ants and termites, which they capture with their long tongues, after foraging in the ant and termite mounds with their strong claws. Like that of other ant- and termite-eaters, the dentition of the aardvarks is greatly reduced (without incisors or canines) and formed by a number of tubular, ever-growing, unrooted premolars and molars (hence the name tubulidentates, coined for this order of mammals).

Another "African" group, the hyraxes, which persisted throughout the Vallesian–Turolian boundary, reached gigantic dimensions in the Greek-Iranian Province (some species of *Pliohyrax* were the size of a tapir).

The Turolian was also the time during which the monkeys of the genus *Mesopithecus* (*M. penteleci*) spread over the Greek-Iranian Province (but not in western Europe), where they replaced the late Vallesian hominoids. These fossil monkeys are included within the subfamily of the colobines, the leaf monkeys, since they shared with them the peculiar cheek-teeth adaptations for browsing leaves. However, the skeleton of *Mesopithecus* indicates a more terrestrial lifestyle.

Among the small mammals, the Turolian rodent faunas followed the trends already established after the Vallesian Crisis—that is, a dominance of murids and ground squirrels and the persistence of some cricetids and glirids. The murids were diversified into at least three genera: *Parapodemus*, *Occitanomys*, and *Huerzelerimys*. Among the cricetids, the lineage of selenodont *Hispanomys* (in the west) and *Byzantinia* (in the east) continued their trend toward increasing size and hypsodonty. Among the cricetines, the sigmodont *Rotundomys* and *Microtocricetus* disappeared, but the lineage of bunodont-brachydont forms leading to the modern hamsters persisted, represented by the genus *Kowalskia*. The dormice representation remained restricted to the living genera *Eliomys*, *Muscardinus*, *Glis*, *Glirulus*, *Dryomys*, and *Myomimus*.

Among the carnivores, the machairodonts persisted into the Turolian, still represented by *Machairodus* (*M. giganteus*) and *Paramachairodus* (*P. ogygia*). *M. giganteus* and other Turolian machairodonts showed further development of the machairodont traits, such as larger, more flattened, curved upper canines and bladelike, serrated teeth (figures 5.21 and 5.22; plate 10). A peculiar group of machairodont cats were the metailurines, represented in the Turolian by a number of genera, such as *Metailurus* (*M. major*, *M. parvulus*) and *Stenailurus* (close to *Metailurus*), the size of a leopard. The skull was characterized by its short face

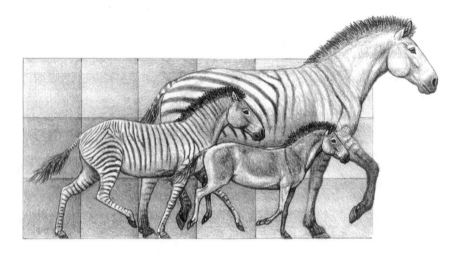

FIGURE 5.17 *Three hipparionine horse species existing sympatrically in the Turolian of Spain* Left to right: Hipparion gromovae, Cremohipparion periafricanum, *and the large* Hippotherium primigenium. Each square measures 50 cm.

and reduced cheek-teeth (loss of the second upper premolar). The most striking feature of this group was the possession of rather "normal," moderate-size upper canines, in contrast to those of the more typical machairodonts. However, the canines were flattened like those of the other machairodonts, and no relationship seems to exist with the modern conical-toothed cats. Interestingly, the Turolian also records the appearance of the first small felids with conical teeth, which were at the

FIGURE 5.18 *Skull, musculature, and reconstructed head of the hipparionine horse* Cremohipparion mediterraneum

Beautifully preserved fossils from Pikermi provide a clear picture of the head of this elegant hipparionine. The muzzle was long, the brain case short, and the orbits placed far back in the skull—giving this animal a thoroughly horselike look. Hipparionines, like most Miocene horses, differed from extant species in having hollows, or fossae, of varying depths and sizes in front of the orbits. Some authors give these features a great systematic value, but their possible function remains a mystery. Their position broadly coincides with the area of insertion of muscles that pull back the nose and upper lip, such as the levator naso-labialis, but the insertion itself remains rather smooth and is not surrounded by prominent ridges, so it is doubtful that an enhanced function of these muscles was the only or even the main cause for such excavations. It has also been suggested that the fossae made room for an enlarged diverticulum, a blind branch of the nasal cavity that remains in vestigial state in modern horses, but the function of the diverticulum in today's animals seems as mysterious as that of the fossae in the fossils. Basal length of skull: 40 cm.

base of the radiation of the modern felines—for example, *Felis attica*, recorded in several Turolian localities from Spain to Greece (figure 5.23). The evolution of modern felids with "normal" conical canines is probably related to the spread of open woodlands. Forest-adapted nimravids and machairodonts developed a strategy of rapid killing when preying on large herbivores, and saber-toothed canines played a major role in this adaptation. Correspondingly, they displayed a long neck and robust anterior limbs, to secure the rapid immobilization and death of the prey. Modern felines are better adapted to prey on large herds of herbivores in the context of open woodlands. The development of more cursorial locomotor adaptations that enabled pursuit and cooperative

FIGURE 5.19 *A scene from the Turolian site of Pikermi, Greece, with two hipparions of the species* Cremohipparion mediterraneum *galloping*

The abundant postcranial remains of this equid found at Pikermi and other sites show that it was an elegant, well-proportioned animal about as tall at the shoulders as a modern Burchell's zebra but considerably lighter in build. Herds of these hipparionines would have moved between the more and the less open woodland. More awkward in appearance were the chalicotheres of the genus *Ancylotherium*, seen in the background as it rears on its hind legs to reach the foliage of a tree with its clawed forefeet. The antelope *Paleoryx pallasi*, with its long, straight horns, would have been somewhat similar to the modern oryx, adding to the "African" air of the Pikermi fauna. Reconstructed shoulder height of *Hippotherium*: 1.2 m.

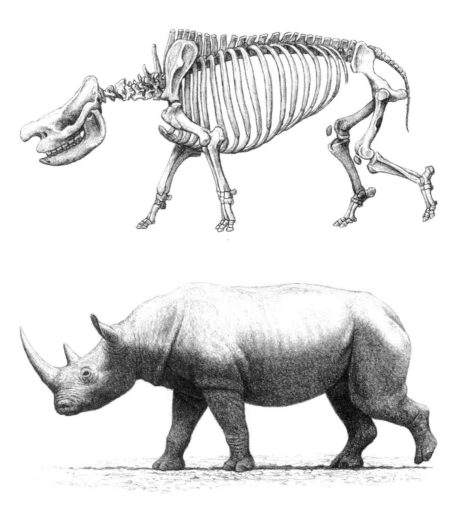

FIGURE 5.20 *Skeleton and reconstructed life appearance of the grazing rhinoceros* Ceratotherium neumayri

The remarkably complete remains of *Ceratotherium neumayri* from Pikermi indicate
that it closely resembled the modern white rhino, *Ceratotherium simum*, the only
surviving species of the genus. The Turolian species showed the features typical of
the genus, such as high-crowned teeth, an elongated head, and heavy limb bones,
but all in a more moderate degree than in extant species. Reconstructed shoulder
height: 1.37 m.

predation probably replaced the mechanism of rapid killing developed
by the robust saber-toothed cats.

Among the hyaenids, there was an extraordinary diversification during Turolian times. Besides *Protictitherium* (*P. crassum*), which in the
Greek-Iranian Province survived the Vallesian Crisis, the wolflike hyaen-

FIGURE 5.21 *Skull, musculature, and reconstructed head of the saber-toothed felid* Machairodus giganteus

Although this species is widespread in many European sites of Turolian age, the best cranial remains actually come from China. The Chinese skulls show a greater machairodont specialization than seen in *Machairodus aphanistus:* the muzzle is more elongated, the mastoid region is projected inferiorly, and the coronoid process in the mandible is reduced, among other adaptations. All these features enabled the cat to open its jaws to around 90° and efficiently bite the neck of large ungulates. Total length of skull: 37 cm.

ids included *Thalassictis, Ictitherium, Hyaenotherium,* and *Hyaenictitherium* (figure 5.5). The cursorial type, represented by *Hyaenictis,* also increased its diversity with *Lycyaena,* which extended its range from Eurasia to northern Africa; remains of this cursorial hyena have been found in Tunisia. But the most common hyaenids from Spain to Iran were the mongooselike *Plioviverrops* and the large bone-cracker *Adcrocuta.* Within the hyena-like guild, the percrocutid *Dinocrocuta* persisted and was found in some localities of Spain (Los Aljezares and Teruel), Greece, and Turkey (Kayadibi and Inönti). However, some kind of competitive exclusion probably existed with *Adcrocuta,* since the two genera very rarely appear together in the same locality.

Indarctos atticus, a species that displayed much larger body size than the first Vallesian representatives of the genus *Indarctos* (about 350 kg, like a brown bear), represented the large ursids. Significantly, the Turolian also records the first entry into Eurasia of the true canids of the genus *Canis* (*C. cipio,* from the middle Turolian of Concud and Los Mansuetos, Spain). The canids had a long history in North America (from the middle Eocene), but did not reach Eurasia until late Miocene times.

SURVIVAL IN AN ISLAND ENVIRONMENT

As we have seen, the hominoids became virtually extinct in Europe after the Vallesian Crisis, pursuing their evolution in southern Asia and Africa. Some hominoids persisted for some time in special refuges in Europe. One of these refuges was the Caucasus, where the presence of a slender dryopithecine (*Udabnopithecus,* actually a small form of *Dryopithecus*) has been reported in the early Turolian beds of Udabno, Georgia. The persistence of evergreen subtropical forests, linked to the retention of special climatic conditions, probably enabled *Udabnopithecus* to survive in this region.

Even more significant was the case of *Oreopithecus,* an enigmatic hominoid that lived in the Tuscany area (northern Italy) from 9 to 7 million years ago. When *Oreopithecus* occupied the Tuscany region, the territories that make up the present Italian peninsula formed an arch of isolated islands that extended from central Europe to northern Africa. One of these islands, close to the European continent, included today's Tuscany and the Corso-Sardinian block. A number of European immigrants settled in this area at a moment between the Vallesian and the Turolian and persisted there until the end of the Miocene. The *Oreopithecus* faunas occurred in several localities from Tuscany, like Casteani, Montebamboli, Ribolla, and Montemassi, as well as Fiume Santo in Sardinia. The best sequence is recorded in the Baccinello Basin, again in Tuscany, where a succession of fossiliferous levels have been recorded.

The lowermost ones, called V0 and V1, are early Turolian in age and still include some "common," nonendemic elements like the cricetid *Kowalskia* and the murids *Huerzelerimys* and *Parapodemus.* These levels also record a number of endemic mammals distinguishable from their ancestors on the continent.

Among the small mammals was *Anthracoglis,* a dormouse close to some Miocene continental representatives like *Microdyromys* and *Bransatoglis* but significantly larger. A second unnamed dormouse is scarcely present in the V1 level of Baccinello. The main feature of this second species was its gigantic body dimensions, larger than those of any extant or fossil glirid. The increase in size observed among the glirids of Tuscany is a common feature among several insular rodents. The French evolutionary biologist Louis Thaler explained this trend to increased size among small insular mammals on the basis of the absence of predators in an island environment. Since the reduced dimensions of an island do not permit the subsistence of a large carnivore, rodents do not need to maintain the small body sizes that enable them to escape predators on the continent. A small body size is an efficient evolutionary strategy for escaping into the undergrowth or for burrowing, but it is also an expensive one, requiring an almost constant intake of food. Small animals have higher relative surface area than larger ones (to decrease size means to increase relative surface with respect to volume). The loss of energy through heat diffusion is, therefore, a problem for any small mammal. But once isolated on an island, in the absence of carnivores small mammals can attain more convenient, larger body sizes, which in some cases can reach gigantic dimensions when compared with those of their continental relatives. This was the case, for instance, with the endemic glirids from Baccinello, and the phenomenon is again observed in several insular Pleistocene faunas (chapter 7).

Besides *Anthracoglis* and the giant unnamed dormouse, a third endemic small mammal, the lagomorph *Paludotona,* also appears in the V1 level. *Paludotona* was an ochotonid whose body dimensions were larger than those of its coeval relatives in Europe. It would seem, therefore, that at the time of deposition of the V1 level this lagomorph had had a long history of isolation on the Tusco-Sardinian island. The most striking feature of *Paludotona* was its archaic dental morphology, which re-

....

FIGURE 5.22 *Reconstructed life appearance of* Machairodus giganteus

In size and proportions, *Machairodus giganteus* was remarkably similar to a modern lion or tiger, but the head was somewhat narrower, with smaller eyes and a more-elongated muzzle. Such differences would become more apparent when the animal was seen more frontally, as shown in this reconstruction. Reconstructed shoulder height: 1.1 m.

FIGURE 5.23 *Skull, musculature, and reconstructed head of the feline* Felis attica

An especially fine skull of this feline comes from the site of Samos, and it displays typically modern morphology and proportions, similar to those of such extant cats as the serval, or the African golden cat. *Felis attica* differed from members of the ancestral felid genus *Pseudaelurus* in details of the dentition, particularly the loss of the second lower premolar and of the posterior cusp of the lower carnassial. Such dental reduction was associated with the shortening of the muzzle, which would have given the living animal a perfectly catlike appearance. Total length of skull: 13 cm.

lates it to some early to middle Miocene ochotonids like *Lagopsis*. But the last *Lagopsis* disappeared from Europe during the middle Miocene, some million years before the deposition of the Baccinello lignites! In fact, *Anthracoglis* presents a similar situation, its dental morphology relating to that of *Microdyromys*, a small dormouse common during the middle Miocene in Europe but almost absent from the late Miocene record.

These data suggest that a previous middle Miocene faunal background already existed on the Tusco-Sardinian island before the late Miocene settlement of *Oreopithecus* and its allies. The presence of an anthracothere in Casteani (equivalent to the V1 level of Baccinello) supports this conclusion. Since the last anthracotheres disappeared from Europe during the early Miocene (chapter 4), an anthracothere's presence in this area can be explained only as a result of an immigration event from Africa or as its persistence as an early Miocene relict, as in the case of *Anthracoglis* and *Paludotona*. In fact, the Tuscany anthracothere is archaic, clearly differing from the advanced late Miocene anthracotherids of northern Africa. The joint evidence of *Anthracoglis, Paludotona,* and the Casteani anthracothere strongly suggests that the remains of a first immigration wave of early Miocene origin were still present in the Tusco-Sardinian island when the *Oreopithecus* fauna entered this region during the late Miocene.

The *Oreopithecus* faunas of the V1 and V2 levels of Baccinello appear as a sort of impoverished, "miniaturized" Vallesian ecosystem. Although already modified by the new insular conditions, most of the large-mammal species of this second, late Miocene immigration wave can be associated with common groups of the late Vallesian or early Turolian European ecosystems, such as hypsodont bovids (*Thyrrenotragus, Maremmia*), giraffids (*Umbrotherium*), *Microstonyx*-like suinae (*Eumaiochoerus*), dryopithecids (*Oreopithecus*), and *Indarctos*-like ursids. *Thyrrenotragus* and *Maremmia* were small bovids with hypsodont dentition. Both forms were once thought to be African immigrants to the area: *Thyrrenotragus* as a neotragine (the tribe that includes dwarf antelopes and gazelles) and *Maremmia* as a precocious alcelaphine (the tribe that includes the African wildebeests, hartebeests, and impalas). However, some of the features that would indicate a relationship with these African groups could have developed independently as specializations linked to insular evolution. Some of the characteristics associated with these two genera, like the short metapodials of *Thyrrenotragus* and the probable ever-growing incisors of *Maremmia*, are also observed in some Pliocene–Pleistocene insular bovids of the western Mediterranean (*Myotragus*) (chapter 7). Most probably, *Thyrrenotragus* and *Maremmia* were advanced insular forms whose origins can be traced to some eastern bovids of the goat group (Caprinae) like *Protoryx* or *Pachytragus*.

A small suid, *Eumaiochoerus,* also appears in the lignites of Baccinello (V2) and in Montebamboli. *Eumaiochoerus* bore a short snout, elongated spatulate upper incisors, and small, chisel-shaped lower tusks. Despite these dental specializations and its small size, other features would indicate the close relationship of this endemic suid to the large *Microstonyx major,* a common species during the Turolian.

The most outstanding member of the Tuscan faunas was the hominoid *Oreopithecus,* which appears abundantly represented in both the V1 and V2 levels of Baccinello. The French paleontologist Paul Gervais first described the remains of this strange primate in 1872. The systematic and phylogenetic position of *Oreopithecus* has always been controversial. This was a medium-size anthropoid of about 25 kg that displayed a short, small cranium and rather high-crowned cheek-teeth. The molars of this anthropoid present several ridges and cuspules that closely match those of some small browsing suids like *Taucanamo* and *Albanohyus.* Because of these specializations, *Oreopithecus* was for a long time considered a cercopithecid. However, its postcranial skeleton, with long arms, suggested a closer relationship to suspensory apes, such as orangutans and chimpanzees. After the finding of a complete skeleton in the mines of Baccinello in the 1950s, the Swiss paleontologist Johannes Hürzeler claimed that *Oreopithecus* was a true hominid and a Miocene link in the lineage leading to hominids. The hominid affinities of *Oreopithecus* were based mainly on its short jaws and reduced canines, features that at that time were considered diagnostic for the hominid family. Moreover, the premolar had two cusps, like those of the hominids, and not one, as with today's apes. After the finding of the Baccinello skeleton, Hürzeler also concluded that *Oreopithecus* was bipedal, since the pelvis was shorter than that of chimpanzees and gorillas and closer to that of the hominids. However, despite the efforts of the Swiss paleontologist, the hominid affinities of the Tuscan anthropoid were finally rejected, its actual affinities remaining obscure for decades.

New analysis of the Baccinello material by a number of researchers in the 1990s reasserted the hominoid character of *Oreopithecus,* which is now thought to be directly related to the Vallesian *Dryopithecus.* The peculiar cranial and dental features of *Oreopithecus* were a consequence of its evolution under insular conditions. Even more, the Spanish paleontologists Meike Köhler and Salvador Moyà (1997) found evidence that *Oreopithecus* probably stood in a bipedal position most of the time, as Hürzeler had suggested. Analysis of its foot and pelvic anatomy seems to reveal that this ape developed a peculiar kind of bipedism, much different from that of the australopithecids. The hallux of *Oreopithecus* was opened and formed an angle of about 100 degrees with the other toes, enabling the foot to act as a tripod so that the animal could maintain an erect posture. In contrast to that of the agile australopithecines, this foot configuration did not enable *Oreopithecus* to develop a fast bi-pedal locomotion. This, of course, was not a problem on the Tuscan island, where truly large predators were absent and the representation of carnivores was restricted to two fish-eating otters (*Paludolutra* and *Thyrrenolutra*) and an ursid related to the European *Indarctos* (probably an omnivorous, mostly vegetarian form).

About 6.5 million years ago, however, a connection with the continent was established, and new herds of European immigrants entered the Tusco-Sardinian area, including large predators such as *Machairodus* and *Metailurus*. Not surprisingly, *Oreopithecus* and the other endemic genera of the Tuscan fauna faced a rapid extinction, unable to resist the competition of the continental newcomers.

The Tusco-Sardinian block was not the only insular territory along the Italian arch during the late Miocene. To the south, on the Adriatic coast, the promontory of Gargano formed at that time an archipelago of small islands that developed, as Tuscany did, their own endemic faunas. These faunas were composed basically of "small mammals," such as hairy hedgehogs, murids, cricetids, and glirids. The term "small mammal" seems inappropriate when applied to the Gargano faunas, since, as insular forms, most of them were quite large. For example, *Deinogaleryx,* a gigantic hedgehog related to the continental echinosoricine *Lantanotherium* (in turn closely related to today's hairy hedgehog, *Echinosorex*), had a body length of about 75 cm and a skull of 25 cm (figure 5.24). It displayed an elongated face with long jaws and a short brain case. In the jaws, the upper incisors were large and vertically implanted. In the lower jaw, the incisors were procumbent and more insectivore-like, being followed by a pair of large lower canines. Although the configuration of the limb bones indicates that *Deinogaleryx* was not a fast-moving animal capable of catching prey, the large body size and dental adaptations of this erinaceid suggest that it was a predator that probably fed on small vertebrates by searching among litter and vegetation. The blunt shape of its cheek-teeth, capable of crushing hard carapaces, also suggests that *Deinogaleryx* supplemented its diet with crabs and other shallow-water crustaceans, as its relative *Echinosorex* does today.

Deinogaleryx was not the only case of gigantism among the "small mammals" of Gargano. *Stertomys* was a giant dormouse that largely exceeded the size of any fossil and living dormice, being comparable only to the giant unnamed species of Baccinello V1. The crown of its molars was complicated, composed of numerous flat and wide ridges. Another peculiar rodent of Gargano was the cricetid *Hattomys,* again much larger than its continental relatives. The dental pattern of this hamster resembled that of the Vallesian *Cricetulodon*. However, several accessory ridges and even extra cusps appeared superimposed on this basic pattern, resulting in a much more complicated morphology. The result repro-

FIGURE 5.24 *Reconstructed life appearance of the giant erinaceid* Deinogaleryx koenigs-valdi, *drawn to scale with a modern European hedgehog*

The appearance of this huge insectivore was probably more similar to that of modern hairy hedgehogs, such as *Echinosorex*, than to the spiny species.

duced quite well that of the Oligocene and early Miocene *Melissiodon* (although they are certainly not related), suggesting that *Hattomys* may have maintained a similar browsing diet.

But the most striking case of rodent evolution in Gargano was that of *Microtia*. Contrary to its generic name, which seems to refer to a microtine rodent, *Microtia* was, in fact, a murid. The members of this lineage show a trend toward developing hypsodont teeth and increasing the number of dental lobes in the first lower molar, as happened later among the Pleistocene microtines. The oldest representatives in the

Gargano fissure infillings still retained a quite basic murid pattern, composed of a succession of three parallel rows of cusps. However, an anterior extra lobe was already present at this early stage of evolution. Throughout the successive populations of *Microtia,* the number of these extra lobes increased in parallel with the genera's increasing hypsodonty and size. Finally, in the youngest members of this lineage, the first lower molar acquired a fully microtoid pattern, with up to six transverse rows.

Not only such "small mammals" as *Deinogaleryx, Stertomys, Hattomys,* and *Microtia,* but also large mammals, like some mustelids and a bizarre ruminant called *Hoplitomeryx,* populated the Gargano archipelago. The appearance of *Hoplitomeryx* must have been impressive, at least in the case of the males, since this archaic ruminant displayed not only a pair of horns over the orbits, as in bovids, but a third, central horn in a more advanced position between the orbits (figure 5.25). Moreover, two lateral, smaller expansions of the supraorbital appendages departed laterally from the sides of the orbits. These appendages seem to have been covered by a sheath like that of bovids. To complete the picture, *Hoplitomeryx* retained a pair of large, daggerlike canines, like those of the moschoids and archaic cervids. Such a combination of characteristics is unique among ruminants and led the Dutch paleontologist Joseph Leinders (1984) to create a new family (Hoplitomerycidae) to accommodate this endemic form. Despite its peculiar cranial morphology, *Hoplitomeryx* has several features, including the retention of daggerlike canines, that indicate that this genus was actually a member of the early radiation of ruminants. It probably derived from a Miocene moschoid like *Amphimoschus* or *Micromeryx* and developed its particular skull appendages independently of the continental bovids and giraffids.

The age and origin of the initial settlement in the Gargano archipelago have been the subject of debate. In contrast to the Tuscan faunas, all the genera in Gargano were of European origin, with no trace of a possible African influence. Most of the endemic animals point to a late Vallesian or Turolian settlement: *Deinogaleryx* (possibly derived from *Lantanotherium*), *Hattomys* (possibly derived from *Cricetulodon* or *Kowalskia*), and *Microtia* (with a tooth morphology resembling that of the late Turolian *Stephanomys,* but most probably related to the early Turolian *Parapodemus*). The complicated dental pattern of the glirid *Stertomys* may indicate an older origin (middle Miocene), but similar complicated dental patterns are observed in some early Turolian dormice from southern and central Europe assigned to the genera *Ramys* and *Vasseuromys.* The most puzzling case is that of *Hoplitomeryx.* As stated, and despite its cranial specializations, this genus must have arisen from a moschoid ruminant like *Amphimoschus* or *Micromeryx. Micromeryx* persisted

FIGURE 5.25 *Reconstructed head of* Hoplitomeryx matthei, *after the serval cranial fragments recovered from the fissure infillings at Gargano, Italy*

in some early Turolian sites from Spain and Germany. However, its derived lower premolar morphology would seem to exclude it from the ancestry of the Gargano ruminant. *Amphimoschus* fits as a plausible ancestor for *Hoplitomeryx,* though, but this would require that the first settlement of Gargano would have occurred during the middle Miocene (the age of the last record of *Amphimoschus* in Europe). Therefore, as with the Tusco-Sardinian faunas, the Gargano faunas were probably the consequence of two different waves of settlement: the first one during the middle Miocene and the second one during the late Vallesian or the Turolian.

THE MESSINIAN CRISIS

Meanwhile, the northeastward movement of the African plate continued during the late Miocene. This led to a progressive restriction of the gateway between the western Mediterranean Sea and the Atlantic Ocean. As today, the Mediterranean probably had a highly deficient water balance, the inflow of Atlantic waters hardly compensating for the intense evaporation. At that time, about 7 million years ago, the Mediterranean was connected to the Atlantic through two gateways. The northern one, known as the Betic Corridor, passed through the Guadalquivir Basin and a number of small basins in southern Spain (Granada, Guadix-Baza, Lorca, and Fortuna). The reliefs associated appeared at that time as an archipelago of small islands bounded by important reef formations. The southern gateway, known as the Rifian Corridor, was in northern Morocco, near present-day Rabat, and consisted of a series of interconnected basins near Meknes, Fez, Taza, Guercif, and Melilla. Both gateways were separated by a emerged area known as the Alboran Microplate (from the small Alboran Island placed to the east of the Gibraltar Strait).

Between 5.8 and 5.6 million years ago, an important sea-level fall, probably related to glaciation and the new growth of the Antarctic (and, perhaps, Arctic) Ice Sheet, restricted for the first time the circulation between the Atlantic and the Mediterranean. This event led to the isolation of most of the marginal basins in the western Mediterranean, including those of the Betic Corridor. As a consequence, they became quickly desiccated, and huge deposits of salt and gypsum replaced the former marine conditions. However, the connection with the Atlantic still existed, and the global sea level rose again 5.6 million years ago. But the end came soon. The tectonic forces through the Betic and, particularly, through the Rifian Corridor led to the final isolation of the Mediterranean (the eastern Mediterranean connection to the Indian Ocean had ended a long time before). In some 1,600 years (a very short time in geological terms), the Mediterranean Sea became completely desiccated, and the two western and eastern basins were each converted into a type of big saline endorreic lake (like the present Dead Sea). From the margins of the surrounding continental areas, several rivers and floods excavated deep canyons and nourished these interior lakes with their waters. As happened with the middle Miocene Paratethys Sea, huge deposits of salt and gypsum filled the deep basins where open-sea conditions had once dominated. Xeric and dry steppe environments extended over the desiccated areas, with the former islands like the Balearics, Corsica, and Sardinia appearing as big mountain massifs in the middle of the desert.

ON THE SHORES OF THE DEAD MEDITERRANEAN

But what happened to the terrestrial faunas on the surrounding continental regions? Logically, a catastrophic event such as the desiccation of the whole Mediterranean Sea should have had very dramatic effects on these faunas. However, this was not the case. In fact, as recorded in a number of key localities, such as Venta del Moro in eastern Spain and Brisighella in Italy, the beginning of the Messinian (the epoch that covers this time span) coincided with an increase in the diversity of the mammalian communities rather than with a biological crisis. In contrast with the previously characteristic open-woodland and grassland Turolian faunas, the beginning of the Messinian brought about the appearance or reappearance of a number of forest dwellers.

Among the large mammals, the tapirs, absent from Europe since the Vallesian times, reappeared. The modern suids of the genus *Propotamochoerus* also reappeared, represented by *P. provincialis,* a species that differed from the Vallesian *P. palaeochoerus* in its large dimensions (about 150 kg). At this time, the giant *Microstonyx major* persisted in central Europe. Other forest dwellers that appeared in the Messinian were the cervids of the genera *Croizetoceros* and *Pliocervus. Croizetoceros* was a rather small cervid, about 1 m high and 60 kg in weight (the size of today's fallow deer, *Dama dama*). *Croizetoceros* differed from the previous Miocene cervids in the possession of complex antlers with three to five short tines. The antlers were rather long and lyrelike, and the tines departed tangentially from the main stem. *Croizetoceros* was, therefore, the first representative of the group of modern cervids having multitined antlers to radiate in the Pliocene epoch. Although the antler morphology of *Pliocervus* was still primitive and close to that of the other Turolian cervids, the shaft was longer, and a second small tine was also present.

Among the bovids, the highly diversified faunas of boselaphines (*Tragoportax*), gazelles (*Gazella*), and small spiral-horned antelopes (*Prostrepsiceros, Ouzoceros, Nisidorcas, Samotragus, Oioceros, Hispanodorcas*) persisted, particularly in the Greek-Iranian Province. However, a new type of bovid, *Parabos,* appeared at this time. *Parabos* was the first member of the group of massive bovids with robust legs and large and relatively short metapodials represented today by buffaloes, bison, and cattle. Like them, it bore a pair of huge horns, which were placed at the rear of the orbits and oriented backward.

The gomphothere proboscideans, rare during the early and middle Turolian, reappeared at the end of the Miocene as *Anancus. A. arvernensis* was a 3-m-tall gomphothere that probably descended from an Asian member of *Tetralophodon* and differed from its close ancestors in a number of significant features (plate 12). While the dentition of *Anancus* was not much different from that of the other tetralophodont gomphotheres, its skull was much shorter than the elongated ("longirostrine") skulls of *Gomphotherium* and *Tetralophodon,* paralleling those of today's elephants. The mandible was also much shorter, having lost the two lower tusks that were present in the early and middle Miocene gomphotheres. The two remaining upper tusks were long and straight, however, reaching 3 m in length. *Anancus* must have looked much more like a recent elephant than like its longirostrine ancestors.

Besides *Anancus,* the zygodont mastodons, rare during the Turolian, reappeared also, represented by more advanced forms like *Mammut.* *Mammut* (not to be confused with the elephantid *Mammuthus,* the mammoth) was the first-known mastodon genus, having been described (Blumenbach 1799) even before Georges Cuvier coined the name *Mastodon.* *Mammut* was a successful group of zygodont mastodons that developed, in parallel, traits similar to those of *Anancus,* particularly the shortening of the face and the reduction and loss of the lower pair of tusks in the

FIGURE 5.26 *Reconstructed life appearance of the mastodon* Mammut borsoni

Zygodont mastodons of the genus *Mammut* are well known in America, where complete skeletons of *Mammut americanum* have been found at several sites of Pleistocene age, but in Europe they have been fragmentary finds. Things changed in 1996, when a partial skeleton of *Mammut borsoni* was found at the site of Milia, Greece. The specimen, which took three years to excavate, includes complete, well-preserved limb bones, a mandible, a maxilla, and a spectacular pair of complete tusks measuring an incredible 4.30 m in length—the longest of any proboscidean ever found in Europe. The size of the limb bones suggests that this individual was about 3.5 m tall at the shoulders, thus being much larger than members of the American species, but even for such a large proboscidean, the huge, almost straight tusks would have looked disproportionately long.

mandible (figure 5.26). The molars were better adapted to a harsh diet than those of *Zygolophodon*, with more sharpened ridges. *Mammut* flourished during the Pliocene and Pleistocene and successfully colonized the Americas, persisting there until the arrival of *Homo sapiens* during the late Pleistocene.

Among the carnivores, the machairodont cats (*Machairodus giganteus, M. kurteni, M. laskaveri, Paramachairodus orientalis*) persisted and maintained a high diversity during the Messinian. So did the metailurines (*Metailurus major, M. parvulus*), which even increased in diversity with the addition of a new genus, *Dinofelis*. This was an eastern Messinian immigrant, found for the first time in early Pliocene deposits in eastern Europe, having attained a wide geographical range from Eurasia and Africa to North America during the Pliocene and Pleistocene. *Dinofelis* was the same size as a modern puma and, as with other metailurines, had moderate-size, flattened canines. Its body proportions were more like those of a forest-dwelling cat than those of a fast runner, with forelimbs having a rather short forearm, as in the jaguar and the leopard. In the Messinian, *Dinofelis* coexisted with the first modern felids with conical teeth of the species *Felis attica*.

But the most significant changes in the carnivore community took place among the ursids and canids. While some Turolian ursids like *Indarctos atticus* persisted into the Messinian, this group was enriched with the appearance of robust, big bearlike forms of the genus *Agriotherium*. According to the remains of this genus found from the early Pliocene in South Africa, *Agriotherium* was in many respects similar to the modern bears, particularly the brown bear (*Ursus arctos*). It probably had similar habits, although *Agriotherium*'s dentition indicates a more carnivorous diet. As with that of several late Turolian suids and cervids, the presence of *Agriotherium* in Messinian deposits indicates the expansion, rather than the retreat, of the forest and woodlands during that time.

After the first entry of canids (the family of dogs, foxes, and wolves) in the middle Turolian with *Canis cipio*, new forms like the American genus *Eucyon* appeared, as recorded at the locality of Brisighella, Italy. But besides *Eucyon*, other members of the dog family appeared and spread successfully during the Messinian. So did the first members of the fox lineage (*Vulpes*), present at the locality of Venta del Moro in Spain, as well as another vulpine form, *Nyctereutes donnezani* (plate 12). Today, the raccoon dog (*Nyctereutes procyonoides*) is a small dog of 20 cm at the withers that resembles the raccoon in the possession of a characteristic black facial mask. Restricted to the woodlands and forested river valleys of eastern Siberia; Manchuria, China; and northern Indochina, during the Pliocene this group of small dogs was abundantly distributed across several Eurasian localities, including western Europe.

In fact, the spread of modern canids like *Vulpes* and *Nyctereutes* may have had dramatic effects on the highly diversified Turolian faunas of doglike hyaenids. Only *Ictitherium, Lycyaena,* and *Thalasictis* persisted from the group of slender hyenas that had dominated the pursuit-carnivore guild during most of the Miocene. Besides these three genera, the late Miocene hyaenid representation remained reduced to the mongoose-like *Plioviverrops* and the large bone-cracker *Adcrocuta.* Outside Europe, the last hyena-like percrocutids persisted in northern Africa (Sahabi), becoming extinct during the set of events associated with the Messinian Crisis. The "modernization" of the carnivore community during the latest Miocene is also evident in other groups such as the viverrids, whose first modern members (*Viverra*) have also been recorded in the Messinian of northern Italy (Baccinello V3).

But the most important result of the Messinian Crisis and the set of events associated with it was the disappearance of the marine barriers that separated the African faunas from the European ones. As a consequence, a number of African immigrants entered Europe, while European ones made the reverse route. This was certainly the case of the first hippopotamuses known in the European record, whose remains, assigned to the genus *Hexaprotodon,* have been recorded in several localities from Spain (Arenas del Rey, El Arquillo, Venta del Moro, and La Portera) and Italy (Casino and Gravitelli). This may also have been the case for some antelopes of African affinity, whose presence in Europe may also be explained as a result of the Messinian Crisis. Some modern tribes of African bovids that appeared in sub-Saharan Africa toward the end of the late Miocene, such as the Reduncini and Hippotragini, have been recorded in parts of western Eurasia. The presence of reduncines (the tribe that includes today's reedbucks) has been reported from Italy (Gravitelli, Sicily) and Turkey. Reduncines also appeared in the Siwaliks, Pakistan, at the top of Dhok Pathan, slightly later than 7 million years ago. But the most impressive and clear evidence of African faunas in western Europe comes from the late Turolian (Messinian) deposits from southern Spain.

About 6 million years ago, the first camels of the genus *Paracamelus* crossed the Betic Corridor and entered the Iberian Peninsula, where they have been found in a number of localities like Venta del Moro near Valencia and Librilla in Murcia. Today's members of the family Camelidae display a peculiar spatial distribution, restricted to Africa and the Near East in the case of camels, and to South America in the case of llamas, vicuñas, and their relatives. But, in fact, this zoogeographic range is rather recent and masks the fact that camels were during most of their history a group of highly diversified North American artiodactyls. They entered Asia and Africa during the late Miocene and from there settled in Iberia. The establishment of the Isthmus of Panama during the early Pliocene enabled this group to colonize South America. Like horses, camels became extinct in North America during the Pleistocene—which explains their recent disjunct distribution. *Paracamelus,* the first member of this family to enter Eurasia, was a large camel displaying most of the features that characterize this group today, bearing a high-crowned, hypsodont dentition, well adapted to the harsh vegetation of the subdesert environments. Like those of modern camels, the limbs of *Paracamelus* had only two fingers, with fused metatarsals and metacarpals. These fingers tended to be highly divergent, to enable the animals to run on the soft, sandy substrate of the desert.

Another curious group that entered the Iberian Peninsula at the same time as camels and other African genera were the macaques of the genus *Macaca,* present at the locality of Almenara-M, in eastern Spain. As we have seen, after the hominoid extinctions associated with the Vallesian Crisis, the colobine monkeys of the genus *Mesopithecus* were common in the Turolian faunas of the Greek-Iranian Province. However, this was the first time that a cercopithecine monkey entered Europe.

But the effects of the African–Iberian exchange are more evident among the rodents. The murids still dominated the late Turolian–Messinian rodent faunas, increasing their generic diversity with several additions: *Apodemus, Rhagapodemus, Stephanomys, Castillomys, Paraethomys,* and others. The two groups of cricetids that survived the Vallesian Crisis persisted: the modern hamsters of the subfamily Cricetinae, represented by *Kowalskia* and *Apocricetus,* and the long-lived selenodont cricetodontines, represented by the larger and more hypsodont genera *Ruscinomys* (which followed *Hispanomys* in the west) and *Byzantinia* (in the east). But during the Messinian Crisis, a number of African rodents settled for the first time on the Iberian Peninsula, including several gerbils of the genera *Protatera, Myocricetodon, Calomyscus,* and others, which were present in a number of localities in southern and eastern Spain. The gerbils are mouselike rodents that today inhabit the desert and steppe environments of Africa and southwestern Asia. In fact, in some late Turolian localities, such as Salobreña, Granada, and Almenara-M, Castellón, gerbils can represent more than 50% of the rodent fauna, thus providing an idea of the extension to the north of the dry conditions prevailing in northern Africa. In any case, the rodent exchange was not one way, and some typical European genera like *Apodemus* (field mouse), *Castillomys, Ruscinomys,* and *Eliomys* (garden dormouse) also entered northern Africa at the same time.

A SECOND CHANCE FOR THE MEDITERRANEAN

After the desiccation of the Mediterranean, the continuous deposition of thick layers of gypsum at the bottom of the basin continued through about 300,000 years. During this time, a tenuous but continuous drainage of saline water from the Atlantic nourished the endorreic basins, entering in the form of small rivers through the Gibraltar threshold, but during that time they were unable to overcome the evaporation rate. But continuous erosion acting in the Gibraltar area had its effect, and at 5 million years ago, a new rise of the global ocean levels precipitated a sudden change. In a catastrophic flood, the waters of the Atlantic abruptly invaded the Mediterranean again. Over thousands of years, gigantic falls decanted their waters into the desiccated basin, restoring full marine conditions and creating the present sea.

On a global scale, the oceanic levels rose about 60 m, covering several areas on the margins of the continents. The old gulfs and canyons developed during the sea-level fall of the Messinian were now inundated by water, forming fiords and estuarine rias. For instance, in the Rhône Valley in France, marine waters reached to the level of the present-day city of Lyon. The Guadalquivir Valley in southwestern Spain was converted into a deep gulf inside the Iberian Peninsula. Shallow waters also covered other small basins in the western Mediterranean, like Roussillon in southeastern France or Empordà and Llobregat close to the city of Barcelona. In contrast to the Parathethys Sea, which ended its history as a sea during the late Miocene, the Mediterranean finally got a second chance. The world of the Miocene was gone, and the beginning of the modern world's conditions had started.

CHAPTER 6
The Pliocene: The End of a World

A T THE BEGINNING OF THE PLIOCENE, THE BIOSPHERE EN-
tered a new phase of climatic optimum. The period between 5
and 3 million years ago appears as a very favorable one, dominated by a warm and humid climate with low-amplitude changes. Temperatures were, in general, higher than today's by about 5°C, while the
annual precipitation in Europe was 400 to 700 mm higher.

Under these warm and humid conditions, the vegetation of western
Europe largely consisted of an evergreen/warm mixed forest dominated
by Taxodiaceae and other warm-temperate trees, similar to the broad-
leaved mixed forest that exists today in eastern China and California.
Biomes of this kind were found, for instance, during the early Pliocene
in northern Spain and southern France. This kind of evergreen subtropical forest even existed during the early Pliocene in Antarctica, which
shows at this time a considerable reduction of its ice cap.

To the south of Europe, a climatic gradient existed at the beginning
of the Pliocene, increasing for temperatures and decreasing for precipitation. In some areas of the southern Mediterranean (for instance,
southern Spain), where the climate was too dry for the presence of
warm temperate trees, the evergreen mixed forest was replaced by a
xerophytic wood/shrub biome in which the xerophytic plants were
dominant. More to the south, in northern Africa, this biome changed
to an open xeric environment dominated by shrubs and grasses.

THE EARLY PLIOCENE FAUNAS

Under these conditions, the early Pliocene mammalian faunas were composed basically of the same taxa that had been common during the latest
Miocene. Among the proboscideans, the long-tusked gomphothere *An-*

ancus and the zygodont mastodon *Mammut* maintained their position as the largest mammals of the early Pliocene forests. Among the ruminants, the last giraffids and the last boselaphine bovids persisted for some time in some early Pliocene localities (La Calera and Sant Onofre, Spain; Montpellier, France). However, the dominant elements were the gazelles (*Gazella*), which became widespread, as well as some slender spiral-horned antelopes such as *Hispanodorcas*. The group of large bovids like *Parabos* persisted as well and even increased their diversity with a new genus, *Alephis,* larger than the former and with lyrelike horns. The cervids of "modern appearance," *Croizetoceros* and *Pliocervus,* evolved new species bearing more-complicated antlers. Among the wild pigs, the medium-size *Propotamochoerus* crossed the Miocene–Pliocene boundary, but the giant suid *Microstonyx* vanished from the fossil record. However, the appearance of the first modern pigs of the species *Sus arvernensis,* ancestor of today's wild boars and pigs (*S. scrofa*), compensated for this loss.

But most of the African animals that entered western Europe during the Messinian Crisis, such as the hippo *Hexaprotodon* and the camel *Paracamelus,* however, did not survive during the Pliocene in this area. However, after its short appearance in the Messinian of Spain, *Paracamelus* persisted in a number of late Pliocene and Pleistocene localities from southeastern Europe and southwestern Asia, such as Sarikol Tepe, Turkey.

Another case of survival across the Miocene–Pliocene boundary were the aardvarks (for example, *Orycteropus depereti*), which were still present in some early Pliocene localities in western Europe such as Perpignan (plate 12).

The Miocene–Pliocene boundary saw a significant decrease in the diversity of the perissodactyls, particularly among the rhinoceroses. All the hornless aceratherines and teleoceratines disappeared from Europe and the eastern Mediterranean. As a consequence, from the earliest Pliocene onward, the hornless hippolike *Brachypotherium* and the horned *Diceros* (the group of the black rhino) and *Ceratotherium* (the group of the white rhino) became restricted to Africa. In Europe, only the lineage initiated with the large *Stephanorhinus schleiermacheri* survived. The early Pliocene representative of this group, *S. megarhinus,* was mainly a large browser, as were most members of the genus. In fact, nearly all the European and western Asian Pliocene rhinos had brachydont cheek-teeth and were browsers. The only rhino with somewhat higher crowns was *S. miguelcrusafonti,* an endemic species from the middle Pliocene locality of Laina, Spain—which suggests an adaptation to the drier Mediterranean conditions of the Iberian Peninsula and southern France.

As did the rhinos, the hipparionine horses decreased in diversity during the early Pliocene. In Europe, *Hipparion crassum,* a big *Hipparion* with large and robust metapodials, represented the genus. A second smaller species defined in the locality of Laina, *H. fissurae,* was much slenderer, with longer metapodials and hypsodont cheek-teeth adapted to grazing. As with the rhino *S. miguelcrusafonti, H. fissurae*'s adaptation to grazing probably indicates the existence at that time in the Iberian Peninsula of drier conditions linked to the Mediterranean climate. In fact, among the perissodactyls, only the tapirs apparently flourished across the Miocene–Pliocene boundary. They largely succeeded in the early Pliocene, being common in several French and Italian localities (*Tapirus arvernensis* from Montpellier, Perpignan, Tuscany, and so on).

The Pliocene also brought a significant decline among the machairodont cats. From the variety of genera existing during the late Miocene (*Machairodus, Paramachairodus, Metailurus*), only the large metailurine *Dinofelis* persisted (plate 12). In their turn, the group of modern felids with conical canines increased their diversity with the appearance of the first members of the *Lynx* group (represented by the species *Lynx issiodorensis*) (figure 6.1).

Among the hyenas, only one Miocene genus, the mongooselike *Plioviverrops,* survived through the Miocene–Pliocene boundary. From the early Pliocene onward, bone-cracker forms of modern appearance, which are considered the ancestors of today's *Crocuta crocuta* (spotted hyena) and *Hyaena hyaena* (striped hyena), commonly represented the hyenas. In the early Miocene, *Pachycrocuta pyrenaica,* first described in the locality of Serrat d'en Vacquer, in the Roussillon Basin (southeastern France), represented the modern hyenas, which probably arose from a form close to the Miocene *Hyaenictitherium.* However, the cursorial lineage was still alive in *Chasmaportetes lunensis,* a very successful species (figures 6.2 and 6.3). The molars and premolars of this hyenid had sharp cutting edges like those of a machairodontine cat or a feline, far from the crushing morphology of the modern members of its family. It also had slender limbs well adapted to running, and, in fact, the limb proportions resembled those of a pursuit carnivore like the cheetah (curiously, they coexisted during most of the Pliocene with the first cheetahs of the species *Acinonyx pardinensis*). *Chasmaportetes*'s geographic range was one of the largest for a hyaenid, from Europe (where it was described as the genus *Euryboas*) to Asia and Africa. In southern Africa, the remains of *Chasmaportetes* appear associated with those of *Australopithecus* in several caves. *Chasmaportetes* even crossed the Bering Strait, becoming the sole hyaenid to settle in North America. There, during the late Miocene and the Pliocene, *Osteoborus,* a group of canids that developed convergent adaptations with the Eurasian hyenas, occupied the bone-cracker guild. The success of *Chasmaportetes* is even more surprising, since foxes (*Vulpes alopecoides*) and raccoon dogs (*Nyctereutes donnezani*) continued to be the dominant generalized mesocarnivores in Europe during the early Pliocene.

FIGURE 6.1 *Reconstructed life appearance of the felid* Lynx issiodorensis

This lynxlike feline is known from a number of European sites of Pliocene age, including excellent cranial material from Saint-Vallier and skull and postcranials from Perrier, both in France. With a shoulder height of about 60 cm, *Lynx issiodorensis* was as large as the largest lynxes living today, but the body proportions differed significantly: the Pliocene species had a relatively larger skull, with a longer muzzle and stronger insertions for masticatory muscles; a longer and more powerful neck; and shorter and more robust limbs. In all these features *L. issiodorensis* resembled typical big cats like the leopard (it was actually as large as a small leopard), and it was probably adapted to taking larger prey than that hunted by the modern lynxes. It is possible, however, that like modern lynxes, *L. issiodorensis* had a fondness for lagomorph meat, and here we show it with a rabbit of the extinct species *Oryctolagus laynensis.*

The bears, or ursids, crossed the Miocene–Pliocene boundary quite successfully. Although the old *Indarctos* failed to survive once again (since the Vallesian times!), the large robust *Agriotherium* persisted well into the Pliocene. And the ursids even increased their diversity with the entry of *Ursus minimus,* the first member of the lineage leading to the

FIGURE 6.2 *Skull, musculature, and reconstructed head of the hyaenid* Chasmaportetes lunensis

Although widespread during the Pliocene in Europe, the so-called hunting hyena was nowhere abundant, and its skeleton remains incompletely known. However, a beautiful skull found at the site of Puebla de Valverde, Spain, provides a more complete picture of this hyaenid's cranial morphology. The cervical vertebrae are based on material from a closely related species, *Chasmaportetes borissiaki,* from Bessarabia, Moldavia. Total length of skull: 26 cm.

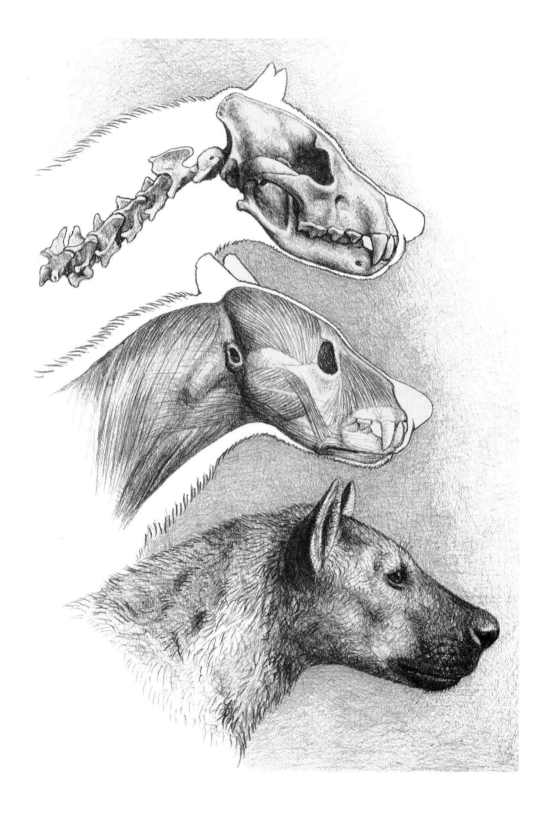

FIGURE 6.3 *Reconstructed life appearance of two hyaenids of the species* Chasmaportetes lunensis

The isolated limb bones of *Chasmaportetes lunensis* from Perrier, France, give an idea of the species's limb proportions, but much additional information for this reconstruction comes from the nearly complete skeleton of a subadult specimen of *C. borissiaki* from Bessarabia. All this material shows that the "hunting hyenas" had slightly longer hind limbs than modern hyaenids, and their heads were not relatively as large. Such proportions were somewhat doglike and are probably related to their active hunting habits, less specialized for scavenging than those of extant true hyenas. Reconstructed shoulder height: 80 cm.

Pleistocene bears and today's *U. arctos* (brown bear) and *U. thibetanus* (Asian black bear). *U. minimus* probably originated in Asia from a Miocene species of *Ursavus* and then migrated into Europe after the extension of the evergreen broad-leaved forests during the early Pliocene.

Among the primates, the cercopithecine monkeys, or macaques, which settled in western Europe during the Messinian Crisis, persisted in this area after the Pliocene flooding. Particularly, the species *Macaca prisca* has been found in several localities in France, Italy, Germany, and Hungary. Although smaller, it may have been conspecific with today's *M. sylvana* from northwestern Africa. Peculiar among the European macaques was *Paradolichopithecus*, a large cercopithecine monkey that developed advanced locomotor adaptations to a terrestrial lifestyle, like modern baboons (it has been suggested that *Paradolichopithecus* may have been the ancestor of the Asian *Procynocephalus*) (figure 6.4).

In addition to the cercopithecines *Macaca* and *Paradolichopithecus*, there was a second group of monkeys, the colobines, or leaf monkeys, which coexisted with the former during the early Pliocene. Among them, *Mesopithecus monspesulanus* was the last member of the Miocene colobine

FIGURE 6.4 *Skull, musculature, and reconstructed head of the cercopithecid monkey* Paradolichopithecus arvernensis

This reconstruction is based on an especially well preserved female skull from Senèze, France. *Paradolichopithecus arvernensis* was a rather large monkey, and in skull morphology it closely resembled modern macaques, while postcranially it displayed more marked terrestrial adaptations, thus resembling extant baboons. Total length of skull: about 18 cm.

monkeys that in the Turolian replaced the last hominoids in Europe. It was mainly a browser and retained an arboreal lifestyle. A second species, *Dolichopithecus rusciniensis,* represented quite a different case because although displaying the colobine dentition adapted to the ingestion of leaves, its skeleton was much better adapted to a fully terrestrial lifestyle, which is an exception among this group of monkeys.

Among rodents, the most diversified family continued to be the murids, with more than six genera, in some cases represented by more than one species: *Apodemus, Rhagapodemus, Stephanomys, Occitanomys, Castillomys, Paraethomys,* and others. The other dominant family was the cricetids, or hamsters, represented by the genera *Ruscinomys* and *Apocricetus.* The gerbils (*Protatera, Debruijnimys*) survived the Messinian Crisis and persisted on the Iberian Peninsula during the early Pliocene. Several other fauna corroborate the warm and humid character of this time. Flying squirrels reappeared after their demise during the Vallesian Crisis as *Pliopetaurista.* Also beavers, after their scarcity during the latest Miocene, were again common in a number of western Mediterranean localities in Spain and southern France (such as Perpignan and Sant Onofre). Their representatives include today's *Castor fiber* or a closely related form.

The climatic amelioration of the early Pliocene affected not only the mammalian communities, but also those of large reptiles. Crocodiles reappeared for the last time along the northern Mediterranean shores of the Guadalquivir Basin, in southern Spain, and giant turtles of the species *Cheirogaster perpiniana,* more than 2 m long, settled in the marshes of the Roussillon Gulf, near the city of Perpignan in southern France (plate 12).

The only interruption to the optimal climatic conditions prevailing in the early Pliocene of Europe was a short cooling peak detected in the isotopic oceanic record at about 4 million years ago (GI16). However, this first discrete cooling event seems not to have strongly affected the composition of early Pliocene terrestrial communities. The only impact that can be tentatively attributed to this shift is the first entry into western Europe of a number of sigmodont cricetids such as *Trilophomys, Celadensia,* and *Bjornkurtenia.* This group of specialized hamsters followed the trend toward developing sigmodont cheek-teeth initiated by some late Vallesian and Turolian genera like *Rotundomys* and *Ischimomys.* In fact, some of them, like *Celadensia* and *Bjornkurtenia,* may be considered primitive arvicolids, or microtine rodents, the family of today's voles and lemmings. The sudden appearance of these sigmodont rodents in Europe can clearly be attributed to a dispersal event of Palearctic origin. These genera were probably high-latitude steppe dwellers that settled in the more southern areas during the rare early Pliocene cooling events that preceded the onset of glacial–interglacial dynamics.

However, the entry of these rodents did not accompany a decline in the highly diversified micromammalian faunas of murids and cricetids, which persisted throughout the early Pliocene.

The first proven arvicolids of the genus *Promimomys* also entered Europe at this time, probably taking part in the same dispersal event from the more northern regions. The immigration of these first arvicolids was a highly significant event, since in this group of rodents the trend toward developing sigmodont teeth reached an extreme. In many respects, the arvicolids are an extraordinary family of rodents. Although their origin is rather recent (the first-known member of the family, *Microtodon,* comes from the late Miocene and early Pliocene beds of Mongolia), today they have a fully Holarctic distribution. The arvicolids possess hypsodont prismatic teeth composed of a number of alternated triangles. During the chewing process, the upper and lower triangles occlude one against the other like the two blades of a pair of scissors; this makes an efficient mechanism with which to graze the hard grains and stems of grasses. With the widespread extension of steppes and prairies associated with glacial dynamics, the arvicolids became the dominant rodents of the Holarctic world. During the early Pliocene, shortly after their entry into Europe, the archaic arvicolids of the genus *Promimomys,* still very close to *Microtodon* and other arvicoloid cricetids like *Celadensia,* were replaced by more-derived forms like *Mimomys* and *Dolomys.* These genera progressed over their ancestors in the possession of more-hypsodont teeth and the addition of a new extra prism in the anterior side of the first lower molars, which, in this way, increased their occlusal surface. Although apparently insignificant, this evolutionary innovation marked the final departure from the basal and archaic cricetid dental design that dominated the evolution of the cricetids and cricetid-like rodents for 30 million years, and enabled the arvicolids to succeed and spread in the hard world of the Pleistocene.

THE FIRST PLIOCENE CRISIS AT 3.2 MILLION YEARS AGO

After the short event at about 4 million years ago, the continuation of the humid and warm conditions can be clearly recognized in the terrestrial record. Outside Europe, the record in eastern Africa indicates that between 3.7 and 3.2 million years ago, there was a "Pliocene Golden Age," with an eastward extension of the rainforest today restricted primarily to the southwest. From 3.7 to 3.4 million years ago, high percentages of mangrove and high river flux were still dominant in the Sahara. Similarly, in northwestern Africa, a humid and probably warm climate prevailed before 3.5 million years ago.

At 3.2 million years ago, however, a first glaciation process developed in the Northern Hemisphere. This glaciation event, which may have involved the development of the first ice sheets in Greenland, led to a strong latitudinal climatic gradient and the first aridification pulse in the Sahara. In eastern Africa, this "cold" event interrupted the "Pliocene Golden Age," as shown at Hadar, Ethiopia, by a much lower proportion of trees (the correlation between "dry" pollen taxa and oxygen isotopes indicates that a link existed between low deep-sea temperatures and drought in northwestern Africa). At this time, the Mediterranean region experienced several more or less minor climatic fluctuations associated with the establishment of the modern Mediterranean climate (with dry and warm summers, cool winters, and humid springs and autumns). Such a succession, with alternating periods dominated by steppe followed by deciduous forests, has been described, for instance, in the preglacial Pliocene sediments of the Villarroya site in Spain (Remy 1958).

In Europe, the climatic event at around 3.2 million years ago involved a main faunal turnover separating the highly diversified faunas of the early Pliocene from those of the late Pliocene. The transition from the early Pliocene (Ruscinian) to the late Pliocene (Villanyian) involved an abrupt decay in small-mammal diversity. The large variety of murids inherited from the early Pliocene suddenly dropped, and *Paraethomys*, *Occitanomys*, and other genera of Miocene origin vanished. At this time, the flying squirrels became rare and almost disappeared from the middle latitudes as well. The cricetid diversity also declined, with the final exit of *Apocricetus* and *Ruscinomys*. The long-lived lineage of the selenodont *Ruscinomys*, initiated some 15 million years earlier, during the early Miocene, finally came to an end. The spread of the arvicolids may have played its role in the extinction of these old hamsters, since, after the first glaciation event, the arvicolids became the dominant small mammals in the whole of Eurasia, with a large variety of genera and species. In Europe, *Mimomys* included several lineages, which developed similar trends in parallel throughout the late Pliocene. These trends included a progressive evolution toward increasing size and hypsodonty, but not the addition of new prisms to the basic arvicolid dental design. As happened during the Miocene with the first hypsodont horses like *Hipparion*, the space between the prisms of the increasingly high-crowned teeth filled with dental cement, to secure a better tooth attachment to the mandible. The evolution toward increasing hypsodonty among the Pliocene arvicolids can be explained in the context of the spread of prairies and steppes, since the presence of silica grains in most of the grasses would have provoked the quick abrasion of teeth and caused rapid extinction (as happened with other groups of rodents).

The cooling at 3.2 million years ago involved a significant turnover among the large mammals as well, and a number of elements of Miocene origin, such as *Propotamochoerus* and the last European giraffids, disappeared. The large bovids of the genus *Parabos* were, in turn, replaced by *Leptobos* (*L. elatus*), a large, robust bovine that resembled the recent genus *Bos* in several features. The powerful horns departed transversally from the rear of the orbits before extending upward. The cheek-teeth were hypsodont, adapted to grazing, and the limbs were robust, with elongated metapodials adapted to running over a hard substrate. The gazellelike group increased its diversity with the entry of *Gazellospira* (*G. torticornis*), a small antelope (about 1 m at the withers) that bore long spiraled horns (figure 6.5). Even more significantly, the ovibovine bovids reappeared at this time in western Europe with *Megalovis* and *Pliotragus*. Both were large bovids (especially *Megalovis*) with hypsodont teeth adapted to grazing. However, they bore relatively small horns compared with their body size. In *Megalovis,* the horns, highly divergent, departed laterally from the front and extended upward. In contrast, *Pliotragus* (*P. ardei*) had rather small horns, which diverged in a V from the rear of the orbits.

The climatic change at 3.2 million years ago also significantly affected the deer, which experienced an important diversification. The late Miocene–early Pliocene *Croizetoceros* evolved into larger species with more-complicated antlers (up to five subdivided prongs in the case of *C. ramosus*) (figure 6.6). Moreover, new cervid species, this time related to today's representatives of the genus *Cervus,* appeared (*C. perrieri* and *C. cusanus*). But the most significant event among cervids was the first appearance of the large megacerine deer. The megacerines were big cervids displaying large antlers. They were common among the Pleistocene faunas and became extinct just 12,000 years ago, together with the large megafauna that characterized the late Pleistocene. *Arvernoceros ardei*, one of the first members of this group, is known from a number of late Pliocene localities, such as Les Etouaires and Vialette, France, and Villarroya and Las Higueruelas, Spain. Although not reaching the giant dimensions common for several Pleistocene megacerines, *A. ardei* was a large deer sharing some of the typical features of the group, such as the possession of long palmate antlers. In the young individuals, a first prong was present at about 10 cm from the rose, the extreme end of the long prong-free antler being anteriorly recurved. In the following years of growth, up to four anteriorly directed tines could appear yearly at the end of the antler, which showed the typical palmate shape of the tribe.

Among the large-browser guild, the two-horned rhinos of the *Stephanorhinus* lineage evolved a new link, *S. elatus* (= *S. jeanvireti*), characterized, like its early Pliocene predecessors, by its large dimensions and robust shape. However, a second, more gracile species, *S. etruscus,* appeared at this time. *S. etruscus* was a relatively small rhino (about 2.5 m long and 1.5 m at the withers), with long, slender limbs, well adapted

to running. This species spread rapidly, from eastern Europe to southern Spain, becoming the most common species in the late Pliocene and
early Pleistocene ecosystems of the continent.

As a consequence of the change in vegetation and in the herbivore
community, the climatic shift at 3.2 million years ago deeply affected
the carnivore guild. Most of the carnivores of "modern" appearance
that had emerged during the early Pliocene persisted, sometimes represented by new species: *Vulpes alopecoides*, *Nyctereutes megamastoides*
(evolved from *N. donnezani*), *Ursus etruscus* (replacing *U. minimus*), *Chasmaportetes lunensis*, and *Pachycrocuta perrieri* (replacing *P. pyrenaica*). How-

FIGURE 6.5 *Reconstruction of the antelope* Gazellospira torticornis, *escaping from the giant
cheetah* Acinonyx pardinensis
Reconstructed shoulder height of *Gazellospira*: 1 m.

ever, a significant renewal took place among the large felids. A number of modern feline genera, such as the first cheetah (*Acinonyx pardinensis*) and the first members of the puma lineage (*Viretailurus schaubi*), are recorded at this time. But the machairodontines also experienced a significant increase in their diversity. Although the old, late Miocene *Dinofelis* disappeared, two successful genera, *Megantereon* and *Homotherium,* spread and joined the diversified large-cat community. Both were widely distributed genera whose remains have been found in several Pliocene and early Pleistocene localities from Africa to China (Nihowan), throughout Spain (Villarroya, Puebla de Valverde, and Guadix-Baza Basin), France (Perrier and Senèze), Italy (Val d'Arno), and other European regions. *Megantereon* was the size of a leopard but, like other machairodontines, differed in its large, laterally flattened canines (figures 6.7 and 6.8). However, the rear of these canines was not serrated,

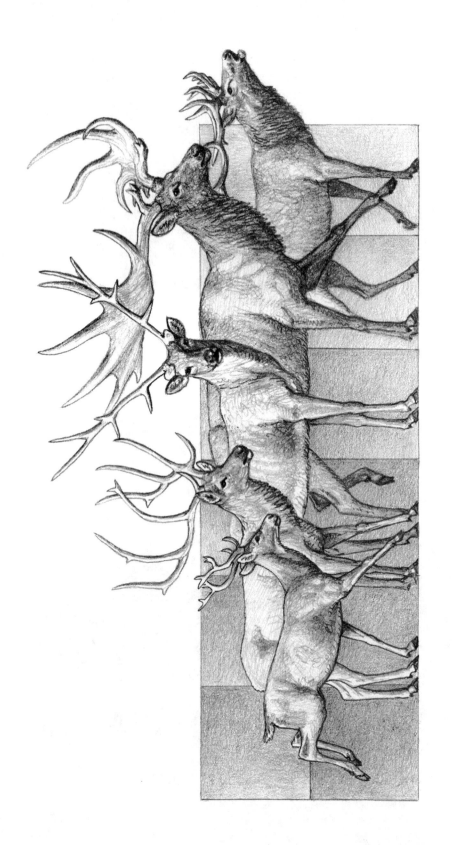

FIGURE 6.6 *Representative species of cervids from the European Pliocene and Pleistocene*

Left to right: Croizetoceros ramosus, Eucladoceros senezensis, Megaloceros savini, Megaloceros giganteus, and *Cervus elaphus.* Each square measures 1 m.

as in most of the machairodontines. It had robust limb proportions closely resembling those of modern jaguars. *Homotherium* was larger (the size of a lion) and had wide and flattened daggerlike, serrated canines (figures 6.9 and 6.10). It had hind limbs of similar length to those of lions and relatively longer forelimbs, which gave the animal a sloping back, vaguely resembling that of a hyena. Both genera crossed the Be-

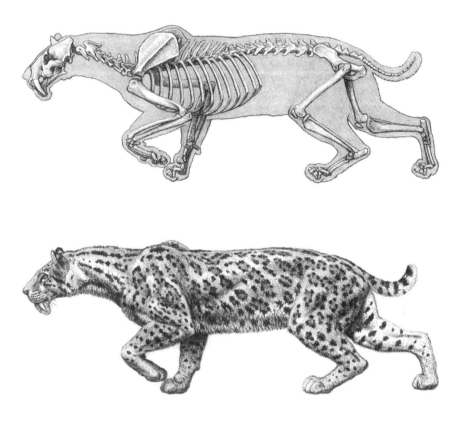

FIGURE 6.7 *Skeleton and reconstructed life appearance of the dirk-toothed cat* Megantereon cultridens

The postcranial anatomy of this species is best known thanks to an almost complete skeleton found at Senèze, which served as a basis for this restoration. Additional information, especially about the back of the skull, was incorporated from material from Valdarno, Italy, and Dmanisi, Georgia. In body proportions and size, *Megantereon* was remarkably similar to the extant jaguar from South America, with shortened, heavily muscled limbs, but there were important differences in the vertebral column: the neck was noticeably longer; the lumbar region, shorter; and the tail, little more than a stub, as in modern lynxes. Such body proportions were related to the particular killing style of dirk-toothed cats, which had to immobilize prey with the strength of their paws and trunk, and then strike with a strong but precise shear-bite to the throat. Reconstructed shoulder height: 75 cm.

FIGURE 6.8 *Two dirk-toothed cats of the species* Megantereon cultridens

The size and body proportions of *Megantereon* indicate that it would have been a good climber and a solitary hunter, and such habits are strongly linked with life in wooded environments. As with modern solitary, forest-dwelling big cats, the only association between adults would occur during courtship and mating. At such times, the encounters between male and female, as depicted in this illustration, would be tense and always bordering on aggression.

ring Strait and entered North America. While *Homotherium* persisted on this continent until the end of the Pleistocene, *Megantereon* evolved into the famous *Smilodon,* known from thousands of skeletons in the late Pleistocene site of Rancho la Brea.

In conclusion, the climatic shift at 3.2 million years ago seems to have been the main agent for the early Villanyian mammal turnover, which caused among other results the entry of several large bovids and cervids like *Leptobos, Megalovis, Pliotragus,* and *Arvernoceros,* and their associated large carnivores: *Pachycrocuta perrieri, Ursus etruscus, Acinonyx pardinensis, Megantereon megantereon, Homotherium crenatidens,* and *Viretailurus schaubi.* The slender anatomy of some ungulates, the spread of grazers, and the diversification of the pursuit carnivores suggest that an extension of the prairies and grasslands took place in Europe as a response to this climatic event.

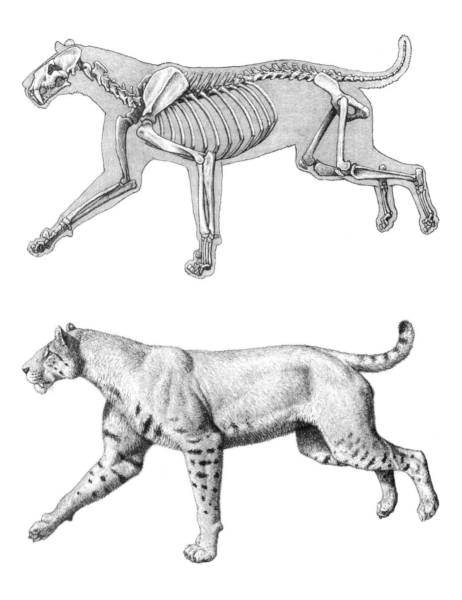

FIGURE 6.9 *Skeleton and reconstructed life appearance of the scimitar-toothed cat* Homotherium latidens

As with *Megantereon,* the most complete skeleton of *Homotherium latidens* comes from the site of Senèze, and it shows that there were considerable differences in body proportions between the two species of sabertooths. *Homotherium* was the size of a modern lion and thus much larger than *Megantereon.* It also had a longer neck than a modern big cat, but proportionally not as long as that of *Megantereon.* The limbs, and especially the forelimbs, of *Homotherium* were proportionally longer, even longer than those of a modern lion, and in some ways a little weaker. Such proportions seem suitable for life in more open environments, and while these cats were obviously adapted to taking large prey, it is possible that group action was necessary to subdue some of the larger prey animals. Reconstructed shoulder height: 1.1 m.

FIGURE 6.10 *Skull and reconstructed head of* Homotherium latidens

There was considerable sexual and individual variation in the morphology of the skull in *Homotherium*. This reconstruction is based mostly on the skull of a large, old male from the site of Incarcal, Spain, and displays an enormous sagittal crest, an almost straight dorsal profile, and a robust mandible. Other, smaller individuals had relatively lower crests, gently convex dorsal outlines, and a more gracile overall build. Basal length of skull: 30 cm.

THE FIRST NORTHERN HEMISPHERE GLACIATIONS AND THE *EQUUS–MAMMUTHUS* EVENT

After the glaciation event of 3.2 million years ago, an important increase in temperature occurred at about 3.1 to 3 million years ago. At this time, temperatures were again about 5°C higher than today's. This event is known as the "Middle Pliocene Warming," and many authors consider it an acceptable analogue of the present-day greenhouse effect. Outside Europe, this warm phase has also been recorded in northwestern Africa, where high percentages of tropical forest indicate the reestablishment of humid conditions between 3.2 and 3.1 million years ago.

But after the Middle Pliocene Warming, the planet finally lost the stable warm and humid climates that characterized the evolution of most of the biosphere during the whole Cenozoic. After hundreds of millions of years, the Earth system entered a new glacial era at 2.6 million years ago that severely affected the continental landmasses. Although the presence of glacier ice-borne deposits has been attested at the high northern latitudes of the Norwegian Sea for more than 6 million years (Jansen and Sjoholm 1991), at this time the first terrestrial ice sheet in the Northern Hemisphere, the Greenland Ice Sheet, extended farther south. At the same time, the Southern Hemisphere saw the growth of the West Antarctic Ice Sheet, which, in contrast with the previous East Antarctic Ice Sheet, was largely sea based.

After this first glaciation event at 2.6 million years ago, the whole planet—and, more specifically, the Northern Hemisphere—entered a new era of periodic extensive glaciations. In contrast with previous Pliocene coolings, this time an actual glacial–interglacial dynamic was established, with alternations of cold- and warm-temperate phases in cycles of 41,000 years. After hundreds of millions of years, the whole of the Earth system entered a cryospheric phase, a cooling situation unknown since the Permian period, during the late Paleozoic.

Although during these first Pliocene glaciations the extension of the glaciers over the continents did not reach the levels attained during the Pleistocene, it strongly influenced the terrestrial climates of the northern and middle latitudes. Therefore, the first glacial phase at 2.6 million years ago (also known as the Praetiglian stage) involved the first replacement of forests by open tundralike vegetation over large areas of northern and central Europe. As a consequence, huge amounts of loess, a sandy windblown sediment produced by the local retreat of glaciers, accumulated in extensive areas to the north of the Eurasian Alpine Belt, from China (where they are hundreds of meters deep) to Europe (where some of the most representative fossiliferous localities, like Saint Vallier in France, are found in this kind of deposit).

The palynological signal for these cold steppe-dominated environments were the dry-adapted wormwood (*Artemisia*), which today colonize both the coldest and the warmest steppes and deserts of the Earth, from the Siberian plains to Death Valley in North America. At the low- and mid-latitudes of the northwestern Mediterranean region, rapid steppe–deciduous forest alternations, linked to glacial–interglacial fluctuations, occurred. The glacial assemblages were similar to those of the modern steppes of the low Mediterranean latitudes that do not have to endure very low winter temperatures and include, besides a large amount of *Artemisia*, a variable proportion of termophyllous taxa like *Cistus* and *Phlomis* (Suc et al. 1995). The interglacial forest assemblages were composed mainly of deciduous elements. In the southwest Mediterranean region (Andalousia, Calabria, Sicily, and northern Africa), *Artemisia* steppe–deciduous forest alternations were not clearly marked, and the only change concerns the increased percentages of *Artemisia* during the latest Pliocene.

The strong development of ice sheets in the Northern Hemisphere around 2.6 million years ago also caused important changes in the vegetation and climate of northwestern Africa that correlate with the establishment of desert and the onset of trade winds. Trade winds increase in periods with large ice volume, which is likewise attributable to an intensification of the high-pressure cell over the Himalayas. Trade winds are connected with sinking airflow and, therefore, contribute to aridity. This is consistent with the pollen succession in northwestern Africa between 2.97 and 2.61 million years ago, which started after 2.8 million years ago with a declining trend in grasses (Poaceae) and was followed by maxima of Choenopodiaceae, Amaranthaceae, and ephedra, and a first maximum of wormwood (*Artemisia*) at 2.6 million years ago. These data point to a strong reduction of the savanna vegetation that was dominant in the early Pliocene and the development of desert in the western Sahara, which at 2.6 million years ago probably reached dimensions close to those of the late Pleistocene (Dupont and Leroy 1995).

To the south, although a few shifts of the desert–savanna boundary (Saharan–Sahelian boundary) were recorded before 3 million years ago, they occurred regularly from 2.6 million years ago onward. At 2.6 million years ago (isotopic stage 104), 2.53 (isotopic stage 100), and 2.49 (isotopic stage 98), severe dry periods are recorded by a high percentage of Choenopodiaceae, Amaranthaceae, ephedra, and wormwood (*Artemisia*). They mark the start of the glacial–interglacial cycles that resulted in arid-cold and humid-warm phases. However, the extension of desert conditions to the south was not a simple question, since the Sahara does not behave as a simple oscillating belt. This is because, while the northernmost position of the front of the southeast monsoon forces oscillations of the southern boundary of the Sahara, changes in the winter precipitation of the Mediterranean climate system determine the northern boundary. As both systems do not oscillate in harmony, the north–south diameter of the desert belt varies in time and space without a stable geographic center.

In eastern Africa, the average percentages of tropical elements also indicate that forests shifted southward after 2.6 million years ago (Bonnefille 1995). In this region, a cooling of 5 to 6°C in the mean annual temperature of the highlands is recorded between 2.51 and 2.35 million years ago in the Gadeb lacustrine diatomite. The colder conditions at Gadeb occurred simultaneously with a strong decrease in tree cover in the Shungura and Koobi Fora Formations. Simulations show that high-latitude glacial ice cover induces cooler winters and drier summers in eastern Africa. Stronger atmospheric circulation would also result from a more extensive northern ice cap, and increased dry northern trade winds would reinforce aridity in the lowlands.

But in spite of their global climatic effects, the onset of continental glaciations in the Northern Hemisphere 2.6 million years ago had limited effects on the middle Pliocene mammalian communities. The European small-mammal associations did not show significant changes comparable to those of the previous Ruscinian–Villanyian boundary at 3.2 million years ago. Similarly, the large-mammal communities remained relatively unaffected. The typical early late Pliocene faunas with *Leptobos elatus, Arvernoceros ardei, Stephanorhinus etruscus, Pachycrocuta perrieri, Ursus etruscus, Acinonyx pardinensis, Megantereon megantereon, Homotherium crenatidens,* and other early Villanyian elements persisted and dominated the terrestrial ecosystems.

Although restricted in terms of ecosystem restructuring, two significant mammalian dispersals were associated with the event of 2.6 million years ago. The first was the entry of the first true elephantids of the genus *Mammuthus,* which, after a period of coexistence, replaced the last gomphotheres of the genus *Anancus.* (*Mammut* had disappeared earlier.) Like today's elephants, *Mammuthus* had a dental pattern completely different from that of the gomphotheres, reflecting its adaptation to grazing grass and other hard vegetation. The cones and tubercules of the mastodon were transformed into a series of small conules that finally became fused in a succession of hypsodont, parallel ridges. The space between the ridges was filled with dental cement, as in other grazers like the horses and voles. The first elephantids probably originated in Africa from an advanced late Miocene gomphotherid species. The oldest member of the group, *Primelephas gomphoterioides,* from the early Pliocene of Chad, retained a pair of small tusks at the end of the mandible. *Mammuthus meridionalis,* one of the first European members

FIGURE 6.11 *Comparison among three species of elephantids from the European Pliocene and Pleistocene*

Left to right: Mammuthus primigenius, Paleoloxodon antiquus, and *Mammuthus meriodionalis.* Each square measures 1 m.

of the lineage, was a large proboscidean, more than 3.7 m high at the withers (figure 6.11). It bore long recurved tusks that diverged from the skull. This species was common in several late Pliocene and early Pleistocene localities of Europe, from Spain and France to Italy, Israel, Georgia, and the Russian plains.

The second main event, shortly after the entry of the first *Mammuthus,* was the replacement of the last *Hipparion* by the first true horses of the genus *Equus.* As at the beginning of the late Miocene, when *Hipparion* replaced the last *Anchitherium,* the history of an American equid replacing the old autochthonous form was repeated. The first *Equus* originated in North America from advanced equids like *Pliohippus,* which increased their size and hypsodonty and reduced the number of toes from three to one. At about 2.6 million years ago (almost coinciding with the boundary between the geomagnetic epochs Gauss and Matuyama), these first representatives of the genus *Equus* dispersed into Eurasia, the oldest ones having been found in Siwaliks, Pakistan, and the Cabriel Basin, Spain. The last hipparionine equids of this time belonged to the species *Hipparion rocinantis,* known from a number of sites in southern Europe (Villarroya, Spain; Perrier-Roccaneyra, France; Kvabebi, Georgia). Curiously, *H. rocinantis* was a large, highly hypsodont and robust form, similar in many respects to the first true horses of the genus *Equus.* This was probably a result of the parallel adaptation to the development of grasslands and prairies over extensive areas of Europe and North America.

Among the first European *Equus* was *Equus stenonis,* which has been found in several localities of western and eastern Europe and included several subspecies (*E. s. stenonis, E. s. livenzovensis, E. s. vireti, E. s. sene-zensis,* and others) (figure 6.12). In general, they were large, quite robust horses between 1.3 and 1.5 m at the withers. Some slender relatives of *E. s. stenonis* (*E. s. granatensis* and *E. altidens*) lie probably close to the origin of the hemions and asses (*E. hydruntinus*), while the origin of African zebras appears to be close to some evolved forms of this species.

Although much rarer than *Equus* and *Mammuthus,* a third immigrant, the large deer of the genus *Eucladoceros,* arrived at the onset of the first northern glaciation. *Eucladoceros* was a big deer, in some species attaining 2 m in length and 1.7 m at the withers (for instance, *E. senezensis*

FIGURE 6.12 *Reconstructed life appearance of the equid* Equus stenonis
Although the stenonine horses belonged in the same genus as all of today's equids, their appearance would have been different because of their large, elongated skulls. This restoration, based mostly on material from the site of Saint-Vallier, shows the long muzzle and concave dorsal profile typical of *Equus stenonis.* Reconstructed shoulder height: 1.4 m.

from the late Pliocene of Senèze, France) (figure 6.6). But the most impressive feature was its large and complicated antlers. These antlers were not palmate, as in the megacerine deer, but ended in four or five long and strong tines, which, in turn, could be subdivided, forming a sort of spectacular comb. *Eucladoceros* persisted into the early Pleistocene in Europe until its final replacement by the large megacerine deer of the genus *Megaloceros*.

However, except for the local extinction of their Pliocene guild equivalents *Anancus* and *Hipparion*, the entry of *Mammuthus, Equus,* and *Eucladoceros* had limited effects on the middle Pliocene European terrestrial fauna. Only the disappearance of some forest-adapted forms like the wild boars (*Sus arvernensis*), the tapirs (*Tapirus arvernensis*), and the colobine monkeys of the genus *Mesopithecus* can be cited as a significant result of the onset of the Northern Hemisphere's glaciation at 2.6 million years ago. It seems, therefore, that the main consequence of this last event was the establishment of new intercontinental bridges between Eurasia and neighboring continents like Africa and North America, leading to the overland dispersal of some significant taxa, such as *Equus* and *Mammuthus*. However, their sudden spread from North America and Africa, respectively, did not cause significant changes in the middle Pliocene's mammal communities, which were probably largely adapted to the further spread of prairies and grasses after the strong faunal turnover at 3.2 million years ago.

In contrast, the glaciation event at 2.6 million years ago severely affected the community of large reptiles, such as crocodiles (which disappeared from the European coasts and marshes) and giant turtles like *Cheirogaster perpiniana*, that had existed during the early and middle Pliocene. These very large chelonians were unable to survive the strong seasonal changes in vegetation cover and temperature that were associated with the glacial dynamics. However, their small relatives of the genus *Testudo* could survive easily (and still survive) because of their ability to dig burrows and hibernate during the cold seasons.

FIGURE 6.13 *Skull and reconstructed head of the canid* Canis etruscus
Abundant fossils of the so-called Etruscan wolf have been known for decades from the Pliocene deposits of Valdarno, Italy, but one of the most beautifully preserved skulls, which served as the basis for this reconstruction, was found recently at the early Pleistocene site of Dmanisi, Georgia. In size and general proportions, the skull is remarkably similar to that of a modern American coyote, displaying the same elongated muzzle and a skull that was in general not only smaller but slenderer than that of true wolves. The external appearance of the head of the living animal would also have closely resembled that of a coyote. Total length of skull: 22 cm.

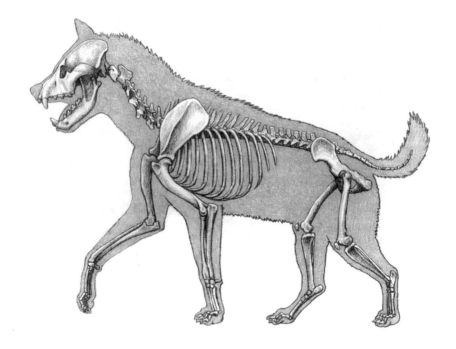

THE WOLF EVENT

The dynamics of mammalian dispersals and faunal turnover initiated with the events of 3.2 million years ago and 2.6 million years ago persisted until the end of the Pliocene. The latest part of the Pliocene (late Villanyian), in particular, saw an important turnover among the carnivores that affected the dog and hyena families (canids and hyaenids). Among the canids, the dominance established by the raccoon dogs (*Nyctereutes megamastoides*) and foxes (*Vulpes alopecoides*) during most of the Pliocene came to an end. While foxes reappeared during the early Pleistocene as *V. praeglacialis*, the raccoon dogs never came back to the continent and became restricted to Asia. In their place, the members of the dog lineage reappeared in Europe, this time represented by the species *Canis etruscus* (figure 6.13). Although smaller than its extant relatives, *C. etruscus* was one of the first members of the lineage leading to today's wolf (*Canis lupus*). *C. etruscus* rapidly became a dominant carnivore in several latest Pliocene and early Pleistocene localities, the dispersal of this species having been named the "Wolf Event" (Azzaroli et al. 1988). A second canid species, *C. (Xenocyon) falconeri*, entered Europe shortly after. This was a large dog that attained the dimensions of present-day northern wolf races. The study of the forelimbs of this species has demonstrated its affinities with the extant wild dog of Africa (*Lycaon pictus*; Rook 1994).

FIGURE 6.14 *Skeleton and reconstructed life appearance of the giant hyena* Pachycrocuta brevirostris

Originally described from a skull found in Seinzelles, France, this hyaenid is known from fragmentary remains from several European sites, but the best sample by far actually comes from the middle Pleistocene site of Zhoukoudian, China. The Chinese sample shows that the body proportions of *Pachycrocuta* especially resembled those of the spotted hyena among modern members of the family, but the extinct species was even more robust. Otherwise, *Pachycrocuta* displayed all the traits that characterize modern hyenas: a large head, a long and well-muscled neck, forelimbs considerably longer than the hind limbs, and a shortened back. All these features are part of an adaptation for carrying large pieces of carcasses and show that *Pachycrocuta* was as adept at scavenging as its extant relatives are. Reconstructed shoulder height: 1 m.

Similarly, the hyenas of this time experienced a significant turnover. *Chasmaportetes lunensis,* the last survivor of the cursorial lineage, finally became extinct, while the slender bone-cracking species *Pachycrocuta perrieri* was replaced by the much more robust *P. brevirostris* (figure 6.14). *P. brevirostris* was a big hyena (about the size of a lion, the largest member of the family), with strong anterior limbs that, as in other modern relatives, were longer than the posterior ones. Its powerful skull was short and high, with massive cheek-teeth well adapted to cracking bones.

Besides *C. etruscus*, *C. (Xenocyon) falconeri*, and *P. brevirostris*, the fourth carnivore to enter Europe during the latest Pliocene was *Panthera gombaszoegensis*. Larger and commoner than the European puma (*Viretailurus schaubi*), this species was directly related to the big-cat group and, more particularly, to today's jaguar (*Panthera onca*).

Among the ungulate world, the entries were much more limited, restricted, in fact, to the first elk of the species *Libralces gallicus* (from the site of Senèze, France), sometimes included in the genus *Cervalces;* the rupicaprine bovid *Gallogoral meneghini;* and one species of the ovibovine bovid *Praeovibos* (whose oldest record comes from the latest Pliocene beds of Almenara 1 and Barranco Conejos, Spain). However, although isolated, these ungulate dispersals were, in fact, highly significant, since they record the first entry of two characteristic elements of the Pleistocene landscapes. On the one hand, *L. gallicus* was the first recognizable member of the lineage leading to today's elk (figure 6.15). Like them, it bore a pair of large palmate antlers that departed laterally from the skull. However, in contrast with those of the modern *Alces alces*, the pedicles of the antlers were long, reaching in some cases 3 m from side to side. On the other hand, *Gallogoral* was rather similar in structure to such present-day rupicaprines as the goral of Asia (figure 6.16). Like the latter, *Gallogoral* was a robust, short-legged antelope, well adapted to locomotion over irregular terrain, but the fossil species was considerably larger than any modern rupicaprine, reaching a height at the withers of 1 m. *Praeovibos,* which replaced the archaic ovibovines of the genera *Megalovis* and *Pliotragus,* was directly related to today's musk ox and was a common dweller on the Pleistocene steppe landscapes.

Even more significant than the entry of *Libralces*, *Gallogoral*, and *Praeovibos* was the disappearance of the slender gazelles like *Gazellospira* and *Gazella* from the continent at this time, the latter becoming restricted to Africa. This event has a high ecological significance, since the distribution of today's gazelles is restricted by the presence of seasonally frozen soils under cold temperate climate conditions.

The large mammalian turnover at the Pliocene–Pleistocene boundary affected a restricted number of species, but was significant because of the main taxa involved. This turnover marked the first entry of the kind of large mammals that would characterize most of the Pleistocene (*Pachycrocuta brevirostris*, *Canis etruscus*, *Panthera gombaszoegensis*, and *Praeovibos* sp.). However, this faunal event had no counterpart among the small-mammal communities. Gradual evolution and local speciation, with no significant dispersal events, marked the late Pliocene evolution of the dominant rodents, the arvicolids. Local lineages of the arvicolid genera *Mimomys* and *Kislangia* tended to develop increasing hypsodonty and cement in the reentrant angles of the cheek-teeth, which has been

FIGURE 6.15 *Reconstructed life appearance of the cervid* Libralces gallicus

The skull, antlers, and much of the skeleton of this spectacular cervid are well known, thanks to fossils found at Senèze. A large deer, *Libralces* stood about 1.5 m at the shoulders, and the widely diverging antlers could spread to 2 m and more. The design of these antlers seems ill-adapted to life in forests, so it appears that members of this species, at least the adult, full-antlered males, would have favored the more open areas among the variety of environments that existed around the site of Senèze.

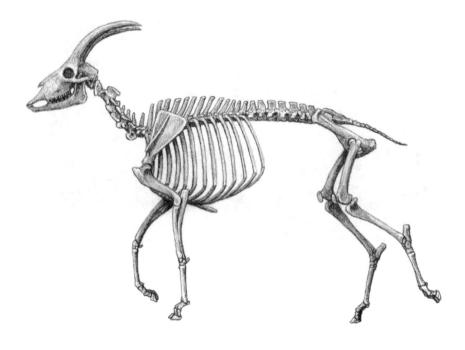

FIGURE 6.16 *Skeleton of the rupicaprine bovid* Gallogoral meneghini
Excellent fossil remains, again from Senèze, show that this animal was rather similar
in structure to such modern rupicaprines as the goral of Asia. Like it, *Gallogoral* was
a robust, short-legged antelope, well adapted to locomotion over irregular terrain,
but the fossil species was considerably larger than any modern rupicaprine, reaching
a shoulder height of 97 cm.

interpreted as an adaptive response to the increasingly cool and arid
conditions.

It is interesting to see how the large- and small-mammal communities
reacted in different ways to the same climate forcing. Not surprisingly,
while dispersal and (local) extinction predominated among the big ungulates and carnivores, the response of the rodents involved much more
local speciation and gradual evolution.

CHAPTER 7
The Pleistocene: The Age of Humankind

THE EVENTS DESCRIBED AT THE END OF CHAPTER 6 WERE, IN fact, the prelude to a new cold phase that started around 1.8 million years ago. At this time, which roughly corresponds to the Olduvai geomagnetic epoch, strong glacial pulses were recorded in the oxygen isotope record of the oceans. At the same time, the pollen record from northern Europe indicates a new extension of the cold steppe, which ended the warm-temperate conditions of the last Pliocene interglacial phase (Tiglian) and led to a new glacial: the Eburonian. This cold event was a global one, recognized in other parts of the world such as eastern Africa, where a marked increase in grasses and other C4 plants reflected a trend toward increasing aridity (Cerling 1992).

THE FAUNAS FROM THE EARLY PLEISTOCENE

The set of cold periods that began at about 1.8 million years ago completed the mammalian renewal that had started at the end of the Pliocene with the "Wolf Event." The most characteristic animal of this new faunal turnover was not a large mammal but an arvicolid rodent, and thus the early Pleistocene across the whole Eurasian continent can be easily identified by the first dispersal of the vole *Allophaiomys pliocaenicus,* from southern Spain to eastern Siberia and even North America. Unlike the Pliocene arvicolids, *Allophaiomys* displayed ever-growing molars in which the roots were always absent. It marked the final step of the evolutionary trend that during the Pliocene had produced progressively hypsodont arvicolids. The last act of this process was the emergence of the voles, in which the hypsodont molars never ended their growth and the roots never developed at the base of the crown. Although some rootless arvicolids were already present among some of the latest Pliocene faunas (*A. deucalion* and *A. vandermeuleni*), the dispersal of *A. plio-*

caenicus was unique because the species attained Holarctic distribution. This dispersal event is significant not only from a biostratigraphic point of view but also from a paleobiological one, since this small microtine was the ancestor of the several vole genera and species that today dominate the subsurface of Holarctic terrestrial ecosystems.

The rapid spread of *A. pliocaenicus* throughout Eurasia can certainly be related to the expansion of the steppe and cooler conditions around 1.8 to 1.6 million years ago, and it was accompanied by a cohort of Palearctic ungulates that joined previously existing large mammals such as *Equus stenonis, Mammuthus meridionalis, Eucladoceros senezensis* (replaced by *E. giulii*), and the carnivores *Canis etruscus, Pachycrocuta brevirostris,* and *Homotherium crenatidens* (figure 7.1). Among them, the ovibovine bovids increased their diversity with the first appearance of *Soergelia,* which joined *Praeovibos. S. minor,* a primitive member of the group found at the site of Venta Micena, was less robust than *Praeovibos* and displayed slenderer horns. The horns, however, were not directed backward, as in most bovids, but were clearly directed forward, thus giving these ruminants a peculiar appearance.

Also significant was the entry at this time of the first members of the genus *Bison* (plate 15). Sometimes included in the separate genus *Eobison,* they were not as large as the middle to late Pleistocene forms, but displayed the features typical of the group. These robust bovids, which were also probably related to the auroch (*Bos primigenius,* the ancestor of domestic cattle), replaced the last *Leptobos* of Pliocene origin (*L. etruscus*) during the early Pleistocene.

Appearing for the first time in Europe were other smaller Palearctic elements, such as *Hemitragus albus,* the first recognizable member of the clade leading to today's *H. jemlahicus,* the tahr from the Himalaya. Like its modern relatives, *H. albus* was a small goat with a robust body and long limbs. Its hypsodont dentition was well adapted to grazing tough vegetation.

Another significant event was the spread of the large megacerine deer, which joined the already large *Eucladoceros. Megaloceros obscurus,* a primitive member of the group, was the first link of the lineage leading to the giant *M. giganteus,* the so-called Irish elk, whose antlers reached more than 3.5 m from side to side and which became extinct at the end of the Pleistocene. A more evolved species, *M. savini,* was common in several early to middle Pleistocene localities, from England to France, Spain, Italy, and Germany (figure 6.6). Although smaller than its late Pleistocene relatives, it bore long antlers that reached 2 m from side to side. These antlers diverged laterally from a broad basal tine, more than 5 cm wide. They were slightly flattened, but without the characteristic palmation of the antlers of the other species of the genus. Taking part in the early Pleistocene cervid community were the members of the fallow deer group. These archaic relatives of the fallow deer

FIGURE 7.1 *Skeleton and reconstructed life appearance of the cervid* Eucladoceros giulii

This reconstruction is based on abundant material from the site of Untermassfeld, Germany, where the postcranials are well represented, giving an accurate idea of the animal's body proportions. Remains of the antlers, however, are fragmentary, and the unknown parts are restored following the pattern of earlier species of the same genus, such as *Eucladoceros senezensis*. Reconstructed shoulder height: 1.55 m.

(*Dama nesti* and *D. vallonetensis*) shared with today's species (*D. dama*) the possession of relatively small palmate antlers (plate 16).

While most of these ungulate immigrants were probably of eastern origin, the early Pleistocene fauna of Europe included at least one African element, the hippo *Hippopotamus major* (figure 7.2). This was a large hippo of about 6 to 7 tons, otherwise very similar to the extant *H. amphibius*. Like today's species, *H. major* bore four incisors (instead of the six present in the Miocene forms) and was mainly a grazer.

But hippos were not the only African group to enter Europe during the early Pleistocene, since recent discoveries in the region of the Caucasus have revealed that this was also the time when early humans dispersed out of Africa. In 1999, the locality of Dmanisi, Georgia, yielded two spectacular skulls of a form close to the species *Homo ergaster* (a mandible had been found in the same site in 1991). The fossiliferous sediments of Dmanisi were deposited over a lava flood that has been dated to 1.8 million years ago. The sediments themselves have revealed a normal magnetic polarity followed by a reverse episode, which indicates an age close to the Olduvai–Matuyama boundary at 1.7 million years ago. The faunal assemblage of Dmanisi is consistent with this age and includes most of the elements that characterized the early Pleistocene faunas of Europe, such as *Equus stenonis*, *Eucladoceros giulii*, *Dama nesti*, *Soergelia* sp., *Canis etruscus*, *Ursus etruscus*, *Pachycrocuta perrieri*, *Panthera* sp., *Megantereon megantereon*, and *Homotherium crenatidens* (Gabunia et al. 2000). However, the presence of some African elements, such as a slender giraffid and a large ostrich, is remarkable. This African influence was also evident among the small mammals, which consisted mainly of gerbils, while the arvicolids were rare.

The two braincases belonged to an adult male and a young female (in which part of the face was also preserved). The most outstanding feature of these two specimens was their low cranial capacity, estimated as 800 cc in the case of the male and 600 cc in the case of the young female. This places the male skull among the lower limits of the cranial

....

FIGURE 7.2 *Skull and reconstructed head of* Hippopotamus major

This reconstruction is based largely on a beautiful complete skull found at the early Pleistocene site of Incarcal, Spain, with additional information from mandibles found in Valdarno, Italy. These Pleistocene hippos were similar to modern members of the genus, but they displayed even more periscopic orbits and a more vertical occiput. The high orbits clearly suggest a more aquatic habit, while the vertical occiput indicates that the neck muscles that elevate the head could rotate the skull along a greater arc. This was an adaptation for the gaping display typical of hippos and suggests that these displays took place more often in the water, where, in order to show its fearsome tusks, the animal had to elevate its head while the mandible remained horizontal, level with the water's surface. Basal length of skull: 70 cm.

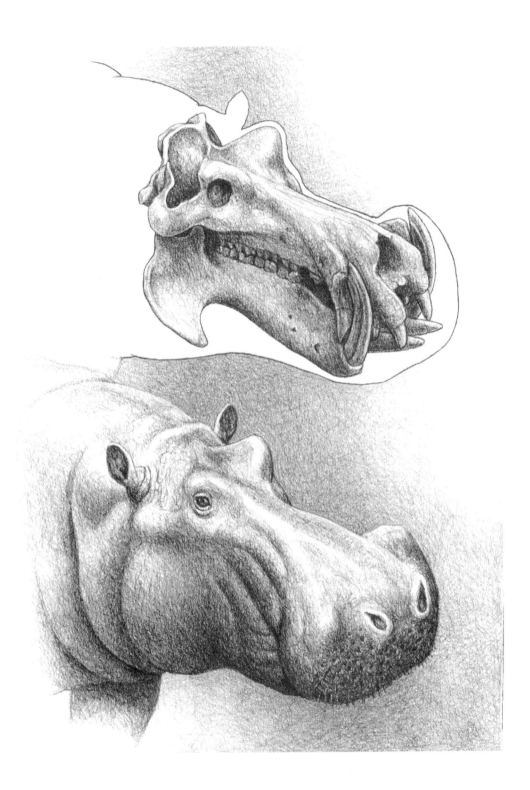

capacity of *Homo ergaster* and related *Homo* species (such as *H. erectus*), while the values of the female specimen clearly lie below those of these species. Moreover, the shape of the two braincases was primitive, with high temporal lines (where the temporal muscles attach to the skull). The individuals from Dmanisi were close in shape and chronology to the most primitive members of the genus *Homo* in eastern Africa. This means that the time between the first appearance of members of the genus *Homo* and their spread out of Africa was almost insignificant.

As with the hippo and the early Pleistocene Palearctic immigrants, this dispersal has to be interpreted within the framework of the climatic and environmental events that took place around 1.8 million years ago. The cold, steppe phase of the early Pleistocene had rather different effects in the Middle East, the area connecting Africa to Europe. For instance, the Eburonian glacial phase corresponded in Israel to a pluvial period and not to a dry phase, and this produced an increase in arboreal pollen in the period between 1.8 and 1.5 million years ago. The glacial phases appeared in this area as cool and humid, with some summer rains, while the dominance of the Saharan regime typified the interglacial periods. This contradictory effect occurred because during glacial times the global climatic belts were to the south of their present position. The belt of western winds was quite active over the eastern Mediterranean, being pushed there by the large quantities of ice over northern Europe, while the Saharan high pressure was much less developed, the tropical region having much less rain. These favorable conditions in the Middle East corridor contrasted with the drier conditions to the south and probably favored the dispersal of early humans into the Middle East.

From the Middle East, *H. ergaster* spread over Asia, where it evolved into the species *H. erectus.* The first members of this species had a cranial capacity comparable to that of *H. ergaster* (about 850 cc) and were present about 1 million years ago as far east as Java (where the original *Pithecanthropus erectus* was described). However, the more-derived populations of the middle Pleistocene in China, about 500,000 years old, reached between 1,100 and 1,200 cc in cranial capacity. The skull of *H. erectus* was long and flattened, with a sharp constriction behind the orbits. The most outstanding feature of this species was its strong superciliary arches, which formed a kind of eyeshade over the orbits. *H. erectus* persisted until the late Pleistocene in Java, where several skulls have been recovered from the terraces of the Solo River. It became extinct after the settlement of the first modern humans in the Far East and Australia.

The End of the Early Pleistocene

After the dry and cold Eburonian sequence, the early Pleistocene ended with a warm phase during which warm interglacial periods alternated with rather mild glacial ones between 1.2 and 0.9 million years ago. Under these conditions, the forests spread again over most of Europe, and a number of temperate elements entered this continent, replacing some of the steppe-adapted forms of the early Pleistocene.

Among the temperate forms, the most characteristic was *Elephas antiquus*, the straight-tusked elephant (figure 6.12). This was a large relative of today's elephants (up to 4 m at the withers), bearing a pair of straight tusks about 2 m long. Throughout the middle and late Pleistocene, *E. antiquus* inhabited the interglacial stages, giving way in the glacial periods to the cold-adapted *Mammuthus*. Meanwhile, *Mammuthus meridionalis* evolved at the end of the early Pleistocene into a number of forms that showed advanced features in their cranial anatomy and dentition (for instance, by increasing the number of ridges, which became narrower). These are usually classified as subspecies of the stem species *Mammuthus meridionalis: M. m. tamanensis, M. m. cromeriensis, M. m. depereti,* and others.

The stenonian horses, derived from *Equus stenonis*, also diversified at the end of the early Pleistocene into a number of species. Some of them were large, such as *E. sussenbornensis*, which stood about 1.8 m at the withers and showed characteristics leading in the direction of the modern horse (*E. caballus*). Others, on the contrary, were small and slender, such as *E. stehlini* and *E. altidens*, the latter probably related to today's hemiones (figure 7.3; plate 14).

Among the rhinos, the old *Stephanorhinus etruscus* evolved into slenderer and more dry-adapted species: *S. hundseimensis*, at first, and *S. hemitoechus* later. *S. hemitoechus* was the size of a black rhino and, like it, bore a pair of long horns (figure 7.4). It had semihypsodont molars probably adapted to grazing grasses and sclerophyllous vegetation. *S. kirchbergensis*, in contrast, was a large rhino. This species could attain 2.5 m at the withers and, unlike the grazers *S. hundseimensis* and *S. hemitoechus*, was a browser with large, low-crowned molars. *S. kirchbergensis* was mainly a forest dweller that characterized the interglacial stages until the beginning of the late Pleistocene.

The appearance of forest, temperate elements, such as the first wild boars (today's *Sus scrofa*) and the macaques (*Macaca sylvana*), attest to the amelioration of the climatic conditions at the end of the early Pleistocene. Among the cervids, the red deer appeared in Europe for the first time, represented by the subspecies *Cervus elaphus acoronatus*. Among the megacerine deer, a giant species, *Megaloceros verticornis*, joined the previous *Megaloceros* species. Like other megacerines, *M. verticornis* had large antlers that departed laterally from the skull, although they differed in being more cylindrical than palmate. *M. verticornis* attained gigantic dimensions for a deer, becoming the largest cervid of its time. Another large cervid was *Alces latifrons*, an elk that could attain

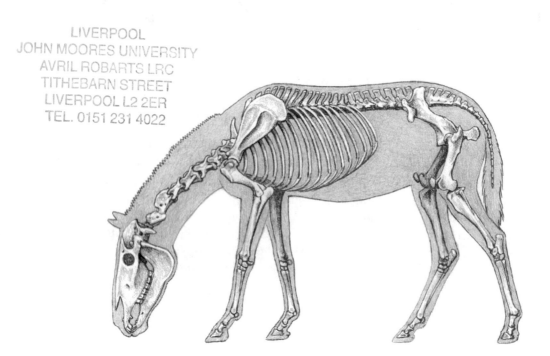

FIGURE 7.3 *Skeleton and reconstructed life appearance of the equid* Equus altidens

This reconstruction is based on a large sample of fossils from the site of Venta Micena, Spain. This material (which some authors classify in a species of its own, *Equus granatensis*) shows all the hemione-like features of the typical *E. altidens:* it was a gracile, long-limbed horse with a relatively big head and slender skeleton, well suited to moving over long distances in relatively open and dry environments. Reconstructed shoulder height: 1.39 m.

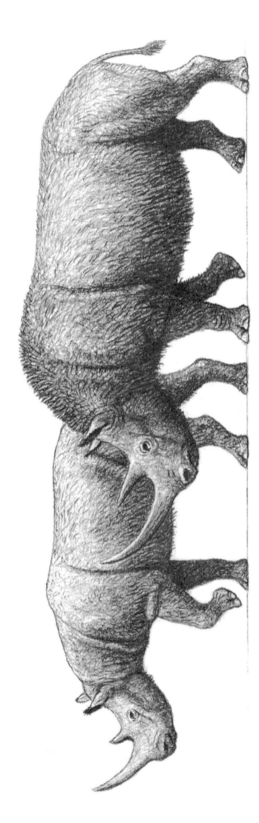

FIGURE 7.4 *Reconstructed life appearance of the rhinoceroses* Stephanorhinus hemitoechus *and* Coelodonta antiquitatis

These two rhino species show increasing adaptations for grazing in open environments. *Stephanorhinus hemitoechus* was a gracile animal (for a rhino) with relatively long legs, and it normally would have carried its head in a moderately inclined position relative to the neck. The woolly rhino (as *Coelodonta antiquitatis* is popularly known) was considerably heavier, with a size and build closely resembling those of the modern white rhino, with which it was not closely related. Details of the soft anatomy of the woolly rhino are well known, thanks to abundant depictions of the animal in cave art, in addition to frozen specimens found in the Arctic permafrost and mummies preserved in petrochemical seeps associated with a salt deposit in Starunia, Poland. Reconstructed shoulder height of *Coelodonta*: 1.7 m.

2 m at the withers. In other respects, it resembled its late Pliocene ancestor *Libralces gallicus,* although the pedicle was much shorter, representing only half the size of the antler and approaching, in this way, that of today's elk (*Alces alces*). Also significant was the first appearance of *Capreolus capreolus,* the roe deer, which today inhabits most of the temperate forests and woodlands of Eurasia, from western Europe to China and Korea. In contrast to most of its Pleistocene relatives, the roe deer is a small forest species bearing rather small antlers with only a few short prongs, which resemble those of the archaic cervids of the Miocene like *Euprox* and *Lucentia.*

The primitive bisons of the genus *Eobison* were replaced by a large species, *Bison manneri,* and a small one, *Bison schoetensacki,* adapted to the new forest conditions. These species displayed rather reduced horns, which measured about 80 cm from side to side and were slightly recurved upward. Another immigrant was *Ovis antiqua,* the first mouflon to enter Europe. Except for some details in the dentition, this species was very similar to today's *Ovis ammon,* the wild sheep that is present in extensive regions of Europe and Asia. Like its modern relatives, *O. antiqua* was probably a grazer preferring the low reliefs and open landscapes instead of the more wooded environments of the late early Pleistocene.

There were also significant changes among the carnivores. From the early Pleistocene *Ursus etruscus,* the Eurasian bears evolved in two directions. An Asian branch maintained an omnivorous diet and relatively moderate dimensions, leading to today's brown bear (*U. arctos*). However, this lineage entered Europe in the late Pleistocene. The other branch developed completely different trends, leading to progressively robust and herbivorous populations. *U. deningeri,* the first member of this lineage, was larger and more robust than *U. etruscus* and was a more vegetarian form, displaying low molars with a large grinding surface. The robust limbs of *U. deningeri* were also an indication of its excavating habits, since it was a hibernating species, like its descendant, the giant *U. spelaeus.* This explains why *U. deningeri*'s remains are frequently found in caves and karstic deposits, which probably were occupied during hibernation.

Among the canids, *Canis etruscus* evolved into *C. mosbachensis,* a more robust and larger form that was, in fact, the direct ancestor of today's wolf (*C. lupus*). A second canid species, *Cuon stehlini,* was also found at the sites of this age. *C. stehlini* was a direct relative of the modern dhole, or Asian wild dog (*C. alpinus*). The dholes are cooperative hunters that inhabit the forests of eastern and central Asia, from India to Java. They are fully carnivorous and differ from wolves in having two molar teeth in each lower jaw, instead of three. Finally, foxes (*Vulpes praeglacialis*) spread again during the early Pleistocene.

Also at the end of the early Pleistocene, the Holarctic *Allophaiomys pliocaenicus* started splitting into a number of independent lineages that were the origin of the modern microtine genera *Microtus, Pitymys, Terricola, Chionomys,* and other voles that inhabit Eurasia and North America. Having exhausted the way of increasing hypsodonty, after the achievement of ever-growing molars, the evolution continued in this group by complicating the rather simple dental pattern of *Allophaiomys*. Thus new extra-fields were added to the archaic, *Mimomys*-like lower first molar of these voles, increasing its shearing surface considerably.

A number of immigrants during the early Pleistocene were of likely African origin. The presence of a tragelaphine bovid (*Pontoceros ambiguus*) has been recorded in Greece (Mygdonia Basin, Macedonia) and Moldavia (Taman Complex). The tragelaphines include large browsing antelopes that today inhabit the woodlands and densely vegetated zones of Africa, such as *Tragelaphus* (bushbucks) and *Taurotragus* (elands). They have spiraled horns and frequently live near wooded areas and even marshes. *Pontoceros* shared with these modern genera the presence of carinate, spiraled horns, and its presence in the Mygdonia Basin agrees with a period of expansion of woodland in large parts of eastern Europe. Other African immigrants settled in Europe during the late early Pleistocene, like the leopard (*Panthera pardus*) and the spotted hyena (*Crocuta crocuta*). The entry of spotted hyenas (which today inhabit extensive areas of Africa) initiated a significant turnover in this group, progressively replacing the large, heavy *Pachycrocuta brevirostris* of the early Pleistocene faunas.

The findings in the cave deposits of Cueva Victoria in southeastern Spain revealed a peculiar case. The faunal association included a typical late early Pleistocene fauna, with *Canis etruscus, Vulpes* sp., *Xenocyon lycaonoides, Panthera gombaszoegensis, Viretailurus schaubi, Homotherium crenatidens, Megantereon cultridens, Pachycrocuta brevirostris, Ursus etruscus, Lynx spelaea, Hippopotamus major, Dama* sp., *Megaloceros savini,* Ovibovini undetermined, *Hemitragus albus, Ovis antiqua, Stephanorhinus etruscus, Equus stenonis, Equus bressanus,* and *Mammuthus meridionalis*. But the most surprising fact was the finding of at least two teeth of the extinct giant baboon, *Theropithecus oswaldi*. *T. oswaldi* was the largest cercopithecid that ever existed, weighing up to 100 kg and reaching more than 1 m at the withers. This giant baboon has been recorded in some late Pliocene localities of northern Africa and coexisted with the australopithecines and early humans in Africa during the early Pleistocene. Its presence in southern Spain is an enigma, since the conditions at 1.4 to 1.2 million years ago seem not to have favored a direct crossing across the Strait of Gibraltar.

Significantly, the first evidence of human occupation in western Europe appears at around 1.2 million years ago. This evidence is not based on direct human remains, as was the case at Dmanisi, but on lithic artifacts found in a number of sites in southern Europe. The oldest ones are Barranco León 5 and Fuentenueva 3 in southern Spain and Le Vallonet in southern France. The sites of Barranco León 5 and Fuentenueva 3 are in the Guadix-Baza Basin and include a typical early Pleistocene assemblage with *Mammuthus meridionalis, Stephanorhinus etruscus, Equus altidens, Hippopotamus major, Hemitragus albus,* and *Soergelia minor.* The sections are composed of reversely magnetized sediments that are older than the normally magnetized Jaramillo epoch. Therefore, the age of the sites extends from the end of the Olduvai epoch (about 1.8 million years ago) to the beginning of the Jaramillo epoch (about 1 million years ago), and the artifacts are probably close to 1.2 million years old (after indirect correlation of the rodent fauna). The site of Le Vallonet seems somewhat younger, corresponding to a cave deposit that delivered a rich mammalian association, including *Ursus deningeri, Cuon stehlini, Canis mosbachensis, Vulpes praeglacialis, Meles meles, Pachycrocuta brevirostris, Homotherium crenatidens, Acinonyx pardinensis, Panthera gombaszoegensis, Panthera pardus, Lynx spelaea, Equus stenonis, Sus* cf. *strozii, Bison schoetensacki, Ovis antiqua, Hemitragus albus, Dama vallonetensis,* and *Megaloceros verticornis.* The age of this assemblage has been established at about 1 million years, according to different methods of dating (basically, paleomagnetism and electronic spin resonance).

THE MIDDLE PLEISTOCENE

About 900,000 years ago, the planet experienced a new, marked decrease in temperature, the first of a number of glacial pulses that were characterized by low temperatures and the extension of the ice sheets over large parts of the Northern Hemisphere (including northern Europe and North America). The climatic cyclicity of 41,000 years, which started in the middle Pliocene, changed again. Now the glacial–interglacial cycles spanned 100,000 years (which seems to conform to the regular changes in the eccentricity of the Earth's orbit), with large glacial stages of about 80,000 years and rather short, although warm interglacial periods of about 20,000 years (we are now living in one of these interglacial times). During its maximal extension in the glacial pulses, the North European Ice Sheet covered all of Scandinavia and part of the United Kingdom, Holland, Germany, Poland, the Baltic countries, and Russia (including Moscow). At the same time, the North Atlantic Ocean was partly covered with ice shelves or pack ice up to the coasts of the Iberian Peninsula and New England. The global sea level dropped in some regions nearly 150 m in response to global glacier accumulation. Consequently, new land bridges and emerged areas broadly connected continental lands such as Eurasia and North America (through the Bering area) or Borneo and Australia. In the middle latitudes, the general scheme of glacial–interglacial vegetation changes was the same as during the previous period: *Artemisia*-steppe during the glacial pulses, alternating with temperate to warm-temperate (depending on the latitude and altitude) deciduous forest during the interglacial phases. In western Europe, the cycles caused the progressive elimination of the termophyllous trees, which persisted as relicts in the eastern Mediterranean region. In northwestern Africa, the pollen record of the past 800,000 years registers latitudinal shifts of up to 10 degrees for desert and wooded grassland/savanna (that is, for the present Saharan–Sahelian boundary).

Under these conditions, most of the large mammals that had appeared at the end of the early Pleistocene persisted, characterizing the faunas that were present in Europe during the warm interglacial pulses: *Elephas antiquus, Stephanorhinus hemitoechus, S. kirchbergensis, Bison schoetensacki, Sus scrofa, Crocuta crocuta,* and *Macaca sylvana.* However, other genera from the early Pleistocene disappeared, such as the large saber-toothed cats of the genus *Megantereon.* These machairodontines survived in Asia until about 500,000 years ago, coexisting with *Homo erectus* (the "sinanthrope") in the famous Chinese site of Zho-Khou-Dien. The saber-toothed cats also persisted and flourished in North America, where a relative of *Megantereon, Smilodon,* reached South America and held on until the end of the last glaciation, 12,000 years ago. In Europe, however, they were replaced by leopards (*Panthera pardus*), lions (*Panthera leo*), and lynxes (*Lynx spelaea*).

With respect to the steppe animals that had characterized the first part of the early Pleistocene, a number of them persisted during the glacial pulses, although normally represented by larger or more robust species. This was the case of the ovibovines *Praeovibos* and *Soergelia* (represented by *P. priscus* and *S. elisabethae*) and the caprines *Hemitragus* and *Ovis* (represented by *H. bonali* and *O. ammon,* today's wild sheep). In the case of *Equus,* the first modern horses related to the present-day *E. caballus* appeared at this time (*E. germanicus* and *E. mosbachensis*) and replaced most of the more slender stenonian species. These middle Pleistocene horses were, in general, larger and more robust than the species of the early Pleistocene and better suited for the cold, steppe landscapes of the middle Pleistocene. Some small stenonian horses persisted into the middle Pleistocene (*E. altidens*) and probably even reached the late Pleistocene (*E. hydruntinus*).

Among the proboscideans, *Mammuthus trogontherii,* the steppe mammoth, which originated somewhere in Central Asia during the early Pleistocene and replaced the last *M. meridionalis* during the middle Pleistocene, characterized the glacial pulses (figure 7.5). *M. trogontherii* was the largest mammoth to settle on the continent, reaching 4.5 m at the withers. Its tusks could attain 5 m and were upward curved. The molars had a more complicated design than those of *M. meridionalis,* with a higher number of straight and plicate ridges.

Joining *M. trogontherii* during the glacial periods of the middle Pleistocene was *Bison priscus* (figure 7.6). This was a robust bovid that resembled today's bison in size and shape, but displayed much longer horns that, in some cases, could attain 1.2 m from side to side. *Bos primigenius,* the wild ox, or auroch, was also common to the middle Pleistocene landscapes. This was a massive species that could attain 2 m at the withers. The males displayed long horns, which were upward recurved at the base, and inward at the tips. The females displayed more slender horns, only upward recurved. Besides *B. priscus* and *B. primigenius,* a third large bovid species, *Bubalus murrensis,* was present among the middle Pleistocene faunas of Europe. In contrast to the former steppe and

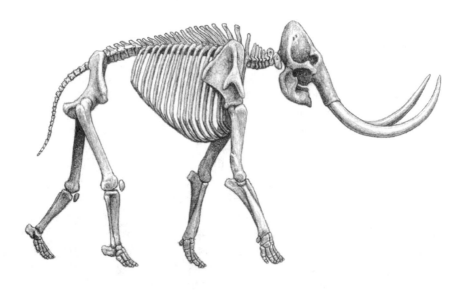

FIGURE 7.5 *Skeleton of the steppe mammoth,* Mammuthus trogontherii
This skeletal reconstruction is based largely on a specimen from the margins of the Dnieper River in Russia that would have attained a shoulder height of about 3 m, but fossils from other sites indicate that the large males in this species could reach a spectacular height of 4.7 m. These measurements make it clear that the steppe mammoth was, on average, considerably larger than the later woolly mammoth, and actually one of the most impressive proboscideans that ever existed. It further differed from the woolly mammoth in having a slightly less sloping back, due in part to its longer hind limbs, and in having a simpler curvature to its tusks.

FIGURE 7.6 *Reconstructed life appearance of the bisons* Bison menneri *and* Bison priscus
The body proportions of the early Pleistocene *Bison menneri* are well known, thanks
to abundant fossils from the site of Untermassfeld that show a tall, long-limbed
animal that would have looked rather different from the later, more typical bisons.
B. priscus, in contrast, was a robust animal with a tall shoulder hump, a feature
related to the habitually low carriage of its head. The proportions in this reconstruction are based on a large sample of *B. priscus* postcranials from Russia. The external
features of this animal, from hair length to fat deposits and even color, are well
known, thanks not only to numerous depictions in the cave paintings of western
Europe, but also to frozen carcasses found in the Arctic permafrost, most notably
the famous "Blue Babe" discovered in Alaska. Shoulder height of *B. priscus:* 1.75 m.

prairie dwellers, *B. murrensis* was, in fact, related to the semiaquatic buffaloes of India (*Bubalus bubalis*) and, like them, probably had a lifestyle
associated with rivers and marshes.

THE HUMAN POPULATIONS
OF THE MIDDLE PLEISTOCENE

A significant event of the middle Pleistocene was signaled by the first
appearance of physical human remains in western Europe, found in
deposits dating to about 800,000 years ago, at Ceprano, Italy, and Gran
Dolina de Atapuerca, Spain. The finding at Ceprano consists of a rather
complete braincase with a cranial capacity of about 1,100 cc. It corresponds to a rather robust hominid with huge superciliary arches over
the orbits that resembles in several features the OH 9 specimen from
the site FLK II of Olduvai Bed II in Tanzania. No faunal remains were
associated with this skull, and its age has been inferred after stratigraphic correlation with other deposits at the Ceprano Basin. The ma-

terial from the karstic complex of Atapuerca comes from the section known as Gran Dolina and was associated with a rich faunal assemblage, including *Canis mosbachensis, Crocuta crocuta, Ursus* sp., *Vulpes praeglacialis, Lynx* sp., *Panthera gombaszoegensis, Stephanorhinus etruscus, Equus* cf. *altidens, Sus scrofa, Dama vallonetensis, Cervus elaphus, Eucladoceros giulii, Bison voigtstedtensis, Castor fiber,* and *Marmota* sp. The paleomagnetic analysis carried out at Gran Dolina indicates that the deposits span the Bruhnes–Matuyama boundary at about 800,000 years ago, although the human remains come from the reversely magnetized part of the section (so they are slightly older than this age).

The human findings consist of a number of cranial remains, including a maxillary and a frontal from a young individual, as well as several mandibular fragments and teeth from the TD 6 level of the section. According to the authors of this finding, the combination of characteristics shown by these remains cannot be fitted into any of the hominid species of this age, so a new species of *Homo, H. antecessor,* was defined (figure 7.7). On the one hand, the mandibular remains indicate a high degree of archaism, with large, not-reduced first and second molars. These archaic features would relate the Gran Dolina species with the former *"H. erectus"* of the Turkana Basin, Kenya, now included in the species *H. ergaster.* On the other hand, according to Bermúdez de Castro and co-workers (1997), the maxillary TD 6-69, although belonging to a young individual, displays an advanced morphology for its age, with some characteristics that are found only in "modern" humans, like those of the site of Djebel Irhoud-1 in Morocco (a locality that is about 600,000 years younger than Gran Dolina of Atapuerca!). A modern shape was also found in the adult maxillary TD 6-58, which presents a marked canine fossa. These advanced characteristics would relate *H. antecessor* to the set of middle Pleistocene hominids found at a number

....

FIGURE 7.7 *Skull, musculature, and reconstructed head of the hominid* Homo antecessor
The site from which this species has been defined is the TD6 level of Gran Dolina de Atapuerca, Spain. This reconstruction is based on combining a partial skull and a right hemimandible (shown here reversed for convenience), both belonging to adolescents, although the mandible likely corresponds to a slightly older individual. Several features of the face contribute to the surprisingly modern appearance of this reconstruction. Some of these features are probably related to the young age of the individual, including the moderate development of the supraorbital torus and the reduced projection of the midface. Other traits are more suggestive of modern humans, including the projection of the nose over the rest of the face and the hollowing of the malar bone under the orbits. Overall, it appears that members of this species would have looked much more familiar to us at around the age of twelve (as shown in this reconstruction) than as adults, when their eyebrows would have become more protruding and their dentition and midface would have projected and given them a more *"Homo erectus*-like" look.

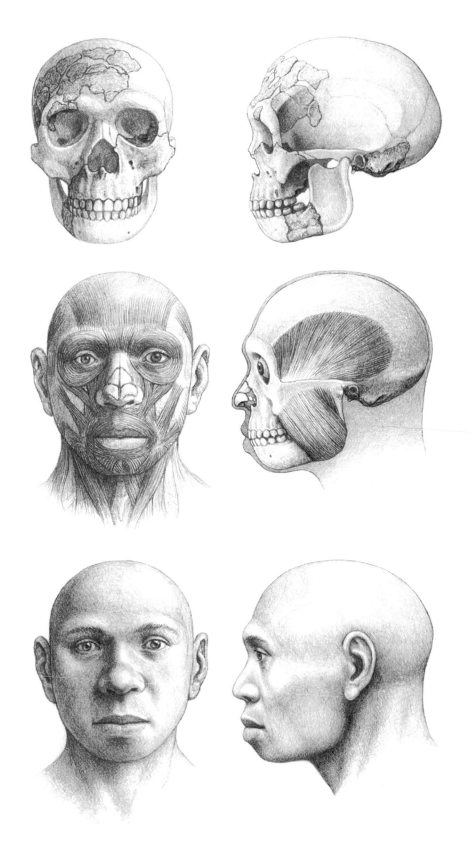

of sites in Europe and assigned to the species *H. heidelbergensis:* Mauer, Germany; Petralona, Greece; Steinheim, Germany; Swanscombe, England; Tautavel, France; and Sima de los Huesos, Spain, again in the Atapuerca karstic complex. Therefore, *H. antecessor,* a direct descendant of *H. ergaster,* would be the ancestor both of *H. heidelbergensis* (which, in turn, led to Neanderthal humans) and of the populations of modern appearance, such as Laetoli 1 and Djebel Irhoud-1, which were present in Africa around 200,000 years ago. In this case, the species *H. erectus* would not have been part of the lineage leading to modern humans or Neanderthals. On the contrary, and in contrast with the point of view that has been maintained for decades, *H. erectus* would have belonged to a separate Asian branch that evolved independently and became extinct without living descendants.

In Europe, at around 600,000 years ago, *H. antecessor* evolved into a more-derived species, *H. heidelbergensis,* whose characteristics directly led to those of the Neanderthals (*H. neanderthalensis*), as indicated by the mandibles found at Mauer and Montmaurin, in Germany and France, respectively, and the partial skulls of Tautavel, Steinheim, Swanscombe, Petralona, and Sima de los Huesos, Atapuerca (figure 7.8). Particularly significant are the findings in this last locality, which has delivered the remains of more than thirty individuals, including males, females, and children. The hominids of Sima de los Huesos are notably younger than those of Gran Dolina, about 320,000 years old. The best-preserved remains from Sima de los Huesos are three partial skulls known as craniums 4, 5, and 6. The cranial capacity of these specimens lies between 1,125 cc (skull 5, probably a female) and 1,390 cc (skull 4, an adult male). These values agree with the other "pre-Neanderthal" specimens assigned to *H. heidelbergensis,* which settled in Europe between 600,000 and 200,000 years ago, like the Swanscombe skull, with an estimated age of 200,000 years and a cranial capacity of 1,325 cc. Another well-preserved specimen, the Steinheim skull, lies within the lower limits of the group, with a cranial capacity of 1,100 cc. This low value, however, may be related to the fact that the Steinheim skull belonged to a female, and this characteristic may have been subjected to strong sexual dimorphism (as in a number of great apes). The cranial capacity of the Tautavel partial skull, with an estimated age of 400,000 years, was also low, probably below 1,200 cc. Altogether, these values were certainly intermediate between those of the primitive *H. ergaster* (between 800 and 1,000 cc) and those of the Neanderthals, which spanned 1,300 to 1,600 cc.

These middle Pleistocene European populations of early humans, although generally scarce, shared a characteristic mixture of archaic and advanced features. The shape of the neurocranium was elongated and closer to those of *H. ergaster* and *H. erectus.* Like them, the front was low and straight and displayed strong superciliary arches. Also, the face and maxillary arches were not as robust and forward directed as in the Neanderthals. However, several specimens also showed a number of features that are closer to the Neanderthal skull morphology, such as larger orbits and wide nasal cavities. In fact, most authors agree that at the end of the middle Pleistocene there was a gradual transition between the youngest *H. heidelbergensis* (such as those of Sima de los

FIGURE 7.8 *Reconstructed life appearance of male and female* Homo heidelbergensis
The anatomy of these "pre-Neanderthals" from the Pleistocene of Europe is well known, thanks to the exceptional sample from the Sima de los Huesos in Atapuerca, Spain. This material shows that *Homo heidelbergensis* was a strong and muscular hominid, but not as short and thickset as the Neanderthals, its likely descendants. As did Neanderthals, both males and females had a wider pelvis than modern humans, implying a wider torso and correspondingly larger body mass. The large sample of *H. heidelbergensis* fossils from Sima de los Huesos shows that the degree of sexual dimorphism in that population was similar to that observed in modern humans. Reconstructed height of male: 1.8 m.

Huesos) and the first Neanderthals, whose first remains are dated at about 200,000 years ago.

THE MEDITERRANEAN ISLANDS: THE LAST REFUGE

The glacial pulses associated with the middle Pleistocene glacial–interglacial dynamics led to important sea-level lowerings in the Mediterranean. As a consequence, a number of islands that had been isolated from the continent were connected to it or became considerably closer to the neighboring landmass. This facilitated the dispersal to these islands of deer, elephants, hippopotamuses, and other groups that were common on the adjacent continent. In the course of the following interglacial pulse, the sea level rose again, and the islands were again disconnected from the continent. The mammals that succeeded in colonizing those islands became, in turn, isolated and pursued their evolution under the new insular conditions. These insular faunas have been found throughout the Mediterranean, in such separated domains as Corsica, Sardinia, Sicily, the Balearics, Malta, Crete, and Cyprus.

Although having independent origins and bearing different faunal elements, all these endemic faunas shared a number of characteristics. They were poorly diversified, usually lacking any trace of large carnivores. This is a logical consequence of the scarcity of resources in the restricted space of an island, which cannot support a great number of large herbivores. Consequently, large predators are usually absent. Both features combined (absence of large carnivores and scarcity of resources) led the surviving large mammals to develop peculiar evolutionary trends, the most outstanding being the evolution toward dwarfism, which in the case of elephants, hippos, and deer reached spectacular results. The small mammals, as we saw for the late Miocene insular faunas, exhibited an opposite trend toward developing large body dimensions.

Another surprising feature of most of these endemic faunas is that, although having originated in different areas and at different times, they were composed of the same elements—that is, deer, hippos, and elephants for large mammals, and mainly dormice for small mammals. Among the typical deer-hippo-elephant insular faunas of the Mediterranean were those of Sicily, Malta, and Crete. In Sicily and on Malta, the evolution toward dwarfism in the elephant group can be followed throughout the middle and late Pleistocene. From the middle Pleistocene *Elephas antiquus*, a first phase of isolation led to *E. falconeri*, about 1 m at the withers. This was a considerable reduction in size, compared with the 3.5 m at the withers of the original straight-tusked elephants. A second isolation phase led to *E. mnaidrensis*, which was 1.5 m or less at the withers! Similar pygmy species of elephants are also recorded in other islands of the Mediterranean, such as *E. melitensis* from Malta (which may belong to the same Sicilian group), *E. cretica* from Crete, and *E. cypriotes* from Cyprus. All of them apparently descended from the same stem species on the continent: the straight-tusked elephant *E. antiquus*.

Another case of spectacular evolution under insularity was that of the Mediterranean hippos. Joining *E. mnaidrensis* in Sicily was *Hippopotamus pentlandi*, derived from the middle Pleistocene *H. major*, but smaller (the size of today's *H. amphibius*). On Malta, *H. pentlandi* evolved into the dwarf *H. melitensis*, the size of a big pig. Another case of dwarfism in hippos was *H. creutzburgi* from Crete, about the size of the so-called pygmy hippo (*Choeropsis liberiensis*) from central Africa (about 1. m at the withers). Even smaller was *Phanourious minor*, also from Crete. The evolution of these dwarf hippos in the context of insularity not simply was a process of shrinking but also involved important skeletal modifications. Comparison of those pygmy hippos with the small (but not dwarf!) *C. liberiensis*, an extant hippo adapted to the canopy forests of central Africa, reveals significant differences in the skull and limb anatomy. Particularly, the distal part of the leg was shortened in the insular forms, with short metapodials and phalanges. This shortening of the distal segments of the limbs is also observed in other insular ungulates, such as dwarf elephants, deer, and goats, and can be regarded as a reversion of the trend toward the lengthening of the metapodials that was present among the cursorial ungulates on the continent. In the restricted space of an island, and in the absence of large predators, large sizes and long limbs are not advantageous features. In contrast, a body close to the ground and a sturdy, solid foot construction would be quite helpful for low-speed locomotion over the uneven topography of an island, which includes both lowlands and mountains (Sondaar 1976).

Similar trends are observed among the dwarf deer that also settled on some islands of the Mediterranean, such as *Megaloceros cazioti* from Corsica and Sardinia. It was a small cervid about 1 m at the shoulders, derived from the big middle Pleistocene *M. verticornis*. The large antlers of the continental species were strongly reduced, and only one or two basal tines were retained. Similar dwarf cervids were present in Sicily and on Malta (*Dama carburangelensis*). An extreme case was *Megaloceros cretensis* from Crete, which was only about 60 cm at the withers and had antlers so reduced that they resembled more those of a muntjac or a roe deer than a megacerine.

The Pleistocene insular faunas of the Mediterranean included not only large mammals but also small ones. In the western Mediterranean were endemic dormice, like those of the Balearics (*Hypnomys, Eivissia*), Corsica and Sardinia (*Tyrrhenoglis*), and Sicily and Malta (*Maltamys, Lei-*

thia). In contrast with the late Miocene insular dormice of the Mediterranean, all these Pleistocene endemics seem to have originated from the same source: the garden dormouse (*Eliomys quercinus*). All, however, were characterized by considerably larger sizes, sometimes double the body dimensions of *Eliomys*. An extreme example was *Leithia melitensis* from Malta, whose skull measured about 10 cm (against the less than 5 cm of *E. quercinus*). In the eastern Mediterranean islands like Crete, similar cases of rodent endemism involving increases in body size were also present (*Kritimys*), although in this case they descended from continental mice (murids of the genus *Mus*, the mouse) and not from dormice. In fact, murids and arvicolids were the most common rodents on the continent when these Mediterranean islands were populated, while dormice were rather rare. Therefore, it is surprising that only dormice represent the rodents among most of the insular faunas in the western Mediterranean. Perhaps the ability of the dormice to hibernate enabled them to survive the cold season, when the resources of an island became scarce.

A peculiar case of insular evolution was that of the endemic fauna of the Balearics, composed of a small goatlike bovid (*Myotragus*), a large dormouse (*Hypnomys*), and a shrew (*Nesiotites*). This association differed from most of the Pleistocene insular faunas in the Mediterranean by its unusual composition, in which no elephant, deer, or hippo was present. The cause for this dissimilarity may lie in the much older origin of the Balearic endemic fauna, which can be traced to the Messinian. In fact, a typical association of *Myotragus*, *Hypnomys*, and *Nesiotites* has been already found in some Pliocene fissure infillings on Mallorca. Most probably, *Myotragus* was an eastern immigrant that settled in the Balearic area during the Messinian Crisis and remained isolated after the Pliocene flooding. Under the new insular conditions, *Myotragus* reduced its size to that of a small goat (0.5 m at the withers) and shortened its metapodials to an extraordinary degree (even more than the dwarf hippos and deer). But the most surprising modifications in *Myotragus* dealt with its skull and dentition. *Myotragus* means "mouse-goat," and this Latin name defines quite well the kind of evolution undergone by this endemic bovid on Mallorca and Menorca. The orbits were positioned in the same frontal plane, as with humans. This advanced position of the orbits permits tridimensional vision and is usually found in arboreal and predatory mammals (such as primates and carnivores), for which estimation of distance is crucial. In contrast, most ungulates have the orbits in a lateral position, which permits a wide range of vision to detect the presence of potential predators. But the most striking modification in *Myotragus* was related to the dentition. The late Pleistocene and Holocene *M. balearicus* bore a highly specialized dentition in which the incisors were reduced to a simple pair in the lower mandible. Moreover, this lower pair of incisors was chisel-like and had no roots, growing continuously, as in rodents. Altogether, the strange combination of small size, shortened metapodials, frontal orbits, and chisel-like incisors made *Myotragus* quite different from the antelope-like ancestor that originally settled the Balearic area in the late Miocene.

Fortunately, the evolution of *Myotragus* since the early Pliocene has been carefully reconstructed after the several fossiliferous fissure infillings that are found on Mallorca and Menorca. The earliest representatives of this lineage (*M. pepgonellae*) still bore an unspecialized dentition close to that of a "normal" bovid, with three rooted incisors plus an incisor-like canine in each lower jaw, and the upper third and fourth premolars were still present. In the following species (*M. antiquus, M. kopperi,* and *M. batei*), a trend toward the loss of the incisor-like canine and the third premolar occurred, while the lower incisors became increasingly hypsodont (although still rooted). Finally, in *M. balearicus* the two lateral incisors and the fourth premolar were lost, and the remaining incisor became unrooted and chisel-like. This gradual process of dental evolution spanned the whole of the Pliocene–Pleistocene, from the early Pliocene *M. pepgonellae* to the late Pleistocene *M. balearicus.* In contrast, the postcranial modifications seen in *Myotragus* occurred just after its isolation during the early Pliocene. *M. pepgonellae,* the first member of the lineage, was as small as *M. balearicus,* had the orbits in a frontal position, and bore extremely shortened metapodials. All in all, *Myotragus* was a strange evolutionary case in which different sets of characteristics evolved at different rates, the skeletal transformations following a punctuated pattern after its isolation in the early Pliocene and the dentition evolving gradually over 5 million years.

THE LATE PLEISTOCENE: THE EXTENSION OF THE STEPPE-TUNDRA BIOME

The late Pleistocene, which includes the last interglacial–glacial cycle, is probably the best-known epoch in the history of the Earth because of the detailed geological and paleontological record that has been preserved from this period in the form of oceanic, glacial, and lake sediments; cave infillings; river terraces; loess; and other kinds of deposits. During the last interglacial pulse, between 130,000 and 115,000 years ago, the climate was, in general, rather similar to that of the present and even somewhat warmer. At the middle latitudes, the summer temperatures were about 2°C higher than today's, and the global sea level was about 4 to 6 m higher. In northern Europe, the glaciers were also less developed than today's, and the vegetation consisted mainly of a broad-leaved/mixed forest dominated by oaks. This phase covers the

Isotopic Stage 5 and is also known as the Eemian interglacial or Riss-Würm interglacial phase (in the old Alpine terminology).

But the conditions changed rapidly, and a sharp cooling started a new glacial phase 115,000 years ago. Between 115,000 and 75,000 years ago, several fluctuations occurred as a succession of stadial (cold) and interstadial (temperate) minor pulses in the context of this glacial phase. Although departing from the glacial conditions of the stadials, the interstadials never reached the warm temperatures characteristic of the interglacial pulses and were much shorter. Between 75,000 and 60,000 years ago, a new rapid cooling took place, and a long cold period started (Isotopic Stage 4). During this time, the glaciers covered most of Scandinavia, and the tundra/prairies carpeted northern and central Europe. But after a new succession of stadials and interstadials (Isotopic Stage 3), the last great mid-latitude ice sheets reached their largest extensions between 21,000 and 17,000 years ago, covering Fennoscandia and the Baltic countries, as well as Poland, north of Germany, Great Britain, and Ireland. The global sea level dropped between 100 and 120 m, and some restricted seas, such as the Mediterranean and Black Seas, were considerably reduced in size (this was not the case of the Caspian Sea, which increased in volume because of the waters coming from the surrounding glaciers). During this cold phase, the temperatures in winter were about 20°C lower than today's, and most of Europe was covered by tundra and cold steppe. The glaciers also overlay the more southerly high reliefs of the Alps, Pyrenees, and Apennines. At this time, only some parts in southern Spain and western Italy were partly forested.

Under these conditions, some of the temperate species that had persisted in Europe during the last interglacial phases came to an end. This was the case of the hippos and the straight-tusked elephant (*Elephas antiquus*). The first cold pulses at about 70,000 years ago led to the extinction of these species and to the spread of *Mammuthus primigenius*, the woolly mammoth (figures 6.12 and 7.9). *M. primigenius* was perfectly adapted to the hard conditions of the steppe-tundra and bore a thick, dense coat that isolated its body from the low temperatures. It was certainly smaller than its ancestors of the middle Pleistocene (*M. trogontherii*), reaching 3 m at the withers. The woolly mammoth displayed long, recurved tusks, which are thought to have helped it dig for roots and tubers in the frozen permafrost of the tundra. The cheek-teeth were composed of a higher number of thinner lamellae (up to thirty on the third molars), which noticeably increased their shearing surface with respect to that of *M. trogontherii* (a clear adaptation to the poor and tough vegetation of the steppe-tundra biome). The limbs and feet of *M. primigenius* had relatively short fur, which shows a good adaptation to a substratum usually frozen and periodically covered by snow. The external appearance of the woolly mammoth is known to us from the several representations of this species that the early modern humans made in several caves of southern France and northern Spain (Font de Gaume, Pech-Merle, and others). However, the best and most direct information about *M. primigenius* comes from a number of complete specimens found in Siberia, in which most of the soft tissues were preserved by the cold conditions in this region.

With the advent of the last glacial pulse, the big, forest-adapted *Stephanorhinus kirchbergensis* was replaced by *Coelodonta antiquitatis,* a two-horned rhino closely related to the *Stephanorhinus* clade but much better suited to the cold and dry conditions of the steppe-tundra (figures 7.4 and 7.10). Like the woolly mammoth, the woolly rhinoceros bore a thick, dense coat that enabled it to resist the permafrost and extreme low temperatures of the last glaciation. Moreover, the cheek-teeth were hypsodont and well adapted to the poor and tough vegetation of the *Artemisia*-steppe. Like those of the woolly mammoth, the limbs and feet of *C. antiquitatis* had relatively short fur. Although the last *S. hemitoechus* persisted in some parts of southern Europe at the beginning of the last glacial cycle, they soon disappeared after the spread of *C. antiquitatis.*

Besides the woolly mammoth and the woolly rhinoceros, other species succeeded on the cold steppe, such as the horse, the bison (*Bison priscus*), and the saiga antelope (*Saiga tatarica*), as well as those animals linked to the present Arctic tundra, such as the reindeer (*Rangifer tarandus*) and the musk ox (*Ovibos moschatus*). The musk ox persisted into the late Pleistocene as the sole ovibovine bovid in Europe, after the extinction of *Praeovibos* and *Soergelia* in the middle Pleistocene (*Praeovibos* probably being within the origin of today's *Ovibos*). *O. moschatus* survives today in the tundra biotopes close to the glaciers of the Nearctic region, from Alaska to Greenland, well prepared for the extremely cold conditions of the Arctic winter, thanks to its dense and long coat.

A similar case is that of the saiga antelope, which still inhabits the steppe environments of Central Asia, from the northern Caucasus to northern China, throughout Kazakstan and Mongolia. This species bears a short trunk, which in the males is much more pronounced than in the females and gives these antelopes a bizarre appearance.

The reindeer, today a migratory species that inhabits the cold forest and tundra of the high latitudes of Eurasia, was also a common dweller of the cold, steppe environments of the late Pleistocene. Other common species from that time that have survived in Europe are the elk (*Alces alces*), the ibex (*Capra ibex*), and the chamois (*Rupricapra rupricapra*). This mixture of steppe and tundra elements also occurs among the small mammals, as reflected by the coexistence of marmots, susliks, and jirds with lemmings. This makes of the late Pleistocene steppe-tundra a peculiar biotope that has no analogue among the present biomes.

The lineage of giant deer of the genus *Megaloceros* also persisted during the last glaciation, attaining its largest dimensions. *M. giganteus,* the so-called Irish elk, was common during the late Pleistocene at the higher and middle latitudes of Europe (figures 7.11 and 7.12). Its robust, largely palmate antlers measured more than 3.5 m from side to side and weighed up to 45 kg. A strong neck and a robust body reaching more than 2 m at the withers supported these enormous antlers, the role of which has been a topic of debate since the nineteenth century, at the very beginning of evolutionary thought. Some paleontologists have shown these antlers, often quoted as a nonadaptive feature, as a case of directed, non-Darwinian evolution—that is, when an organ kept growing and growing independently of the environmental conditions, thus leading to the final (and unavoidable) extinction of the lineage.

In a classic study, done in the late 1960s, Stephen Jay Gould showed that the giant antlers of *M. giganteus* could be interpreted as a case of allometric evolution. When he analyzed the joint evolution of body and antler size in this group, he realized that there was an allometric relationship between the growth rates of these two characteristics. As the body size grew, the antlers increased their size at a higher rate. *M. giganteus* having the largest dimensions among its clade, the size of the antlers was what could be expected according to its giant size. The large antlers of *M. giganteus* were, therefore, interpreted as a consequence of the large body dimensions attained by this species and not as a case of "directed" evolution. Moreover, these enormous antlers probably had a clear display function, preventing most of the fights that are usual among the males of other deer species. This would have involved a considerable economy of energy and prevented the fatal casualties that sometimes occur during these fights. Therefore, although the large antlers of *M. giganteus* could have prevented them from occupying the dense forest where extant deer live, they could have had a high adaptive

FIGURE 7.9 *Reconstructed life appearance of the woolly mammoth,* Mammuthus primigenius

Thanks to numerous skeletons from Eurasia and North America, plus several frozen carcasses preserved in the Arctic permafrost and abundant depictions in cave art, the woolly mammoth is perhaps the best-known representative of the extinct Ice Age faunas of the Northern Hemisphere. This reconstruction shows a male covered in a thick, black winter coat, with robust, strongly curved tusks. In the popular mind, the woolly mammoth is associated with snow-covered landscapes, and snow would indeed have been present in the mammoths' habitat for part of the year, but the "mammoth steppe," as the habitat of the Palearctic Ice Age megafauna is often termed, received much less snowfall than modern tundra does, being overall drier and grassier. This scene is set in the margins of a partly frozen river, with a small wood of larch in the mid-distance. Reconstructed shoulder height: 3 m.

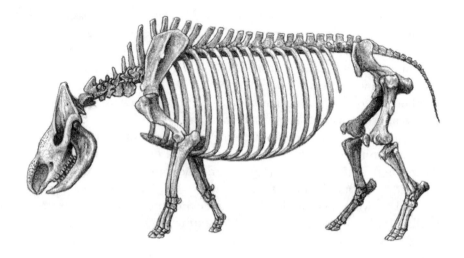

FIGURE 7.10 *Skeleton of the woolly rhino,* Coelodonta antiquitatis

The woolly rhino is known from abundant skeletal remains from many sites in Europe. The body proportions of *Coelodonta antiquitatis* were similar to those of the modern white rhino, as a result of convergent adaptations for efficiently grazing short grasses. As in the white rhino, the head of *C. antiquitatis* was long, and it would normally have been carried in a low position. The limb bones were robust and heavy. Reconstructed shoulder height: 1.7 m.

value as display ornaments preventing the loss of energy and other consequences of seasonal male combats.

Among the carnivores, most of the middle Pleistocene species survived into the late Pleistocene, although represented by larger and more robust "cave" variants that are often described as separate species or subspecies (figure 7.13). Among the hyaenids, the spotted hyena persisted, although represented by its own "cave" subspecies, *Crocuta crocuta spelaea,* which was certainly more robust than its modern relatives. This

FIGURE 7.11 *Reconstructed life appearance of the giant deer* Megaloceros giganteus antecedens

Beautifully preserved remains of this deer, including complete antlers, have been found at the site of Steinheim an der Murr, Germany. Although these animals are thought to belong to the same species as the giant deer of the late Pleistocene, the shape of the antlers is different, with the brow tine expanded into a circular horizontal plane, the back tine subsumed within palmation, and the palmation itself more laterally compressed and upwardly directed. With their less extreme lateral expansion, the antlers of *Megaloceros giganteus antecedens* would have allowed the stags of this species freer movement between meadows and open woodlands than was possible for the later subspecies.

was also true of the cave lion, *Panthera spelaea,* which is usually considered a separate species, differing from today's *Panthera leo.* Certainly, the cave lion was much bigger than today's lions, reaching 2 m in length and 1.2 m at the withers. We know, thanks to the representation of this species by early modern humans, that the cave lions did not display the typical male coat of the African lions, although they did bear the tuft at the end of the tail. They probably looked much more like a female lion, but with a denser and thicker coat (especially in the neck region) and with shorter and more robust limbs.

An extreme example in this evolution toward increasing size and robustness among the late Pleistocene carnivores was the cave bear, *Ursus spelaeus* (figures 7.14 and 7.15). This species took to an extreme the trends already observed in *U. deningeri.* Like its middle Pleistocene an-

FIGURE 7.12 *Reconstructed life appearance of the giant deer* Megaloceros giganteus hibernicus

This giant deer subspecies, popularly (and wrongly) known as the Irish elk, displayed the most spectacular antlers of any cervid, living or fossil, spanning up to 3.5 m. The anatomy of this animal is well known, thanks mostly to beautifully preserved skeletons found during the nineteenth century in peat bogs in Ireland. As in today's cervids, the antlers were shed each year, and a new and larger pair grew each summer. The specimen depicted here is just dropping the velvety coat of skin that covers new antlers during their growth.

FIGURE 7.13 *Selection of mammalian species from the late Pleistocene site of the cave of Lezetxiki, Spain*

The species represented in this illustration are from among those that no longer exist in the Iberian Peninsula, but the fauna was considerably more varied, including modern species such as the red deer, wild boar, chamois, and wolf. The species depicted from left to right are leopard (*Panthera pardus*), cave hyena (*Crocuta crocuta spelaea*), wolverine (*Gulo gulo*), cave lion (*Panthera leo spelaea*), horse (*Equus caballus*), reindeer (*Rangifer tarandus*), bison (*Bison priscus*), and woolly rhino (*Coelodonta antiquitatis*). Each square measures 1 m.

FIGURE 7.14 *Skull and reconstructed head of the cave bear* Ursus spelaeus

Based on specimens from the cave of El Reguerillo, in Spain, this reconstruction shows the traits of late Pleistocene cave bears. The forehead was distinctly domed, and the muzzle was shorter than that of modern brown bears; the nasals were somewhat retracted, and the mandible was robust and displayed a distinct curvature to its lower outline. These features were related to the powerful masticatory apparatus, an adaptation to a diet that included less meat and more tough vegetable matter than that of brown bears. Total length of skull: 45 cm.

FIGURE 7.15 *Reconstructed life appearance of the cave bear, shown in a mountain environment*
In life, the cave bear would have looked considerably more thickset than a modern brown bear. The limbs were more robust, with the distal segments shortened, and the head normally would have been carried rather low. As its name suggests, the remains of the cave bear are normally found in caves, and often in great numbers. The proportions of this particular reconstruction are based on a large sample of fossils from the cave of Arrikrutz, in Spain. Reconstructed shoulder height: 1.2 m.

cestor, *U. spelaeus* was mainly a vegetarian species, bearing even larger molars adapted to grinding fruits and other vegetables. During the last glacial pulse, it attained its largest dimensions, being 2 m long and reaching 1.2 m at the withers. The body was robust, ending in a big and short skull that displayed an elevated front. The name cave bear, by which it is popularly known, seems really appropriate, since its existence appears clearly associated with caves, in which this species hibernated in large numbers during the hard winters of the last glacial period. In fact, there is a huge record of *U. spelaeus* in the form of thousands of skeletons found in several caves in western and eastern Europe. These finds have given us an accurate understanding not only of the physical characteristics of this bear, but also of some aspects of its biology and demography. We know that several young and old individuals died during hibernation in the cold season, probably because of diseases or a dearth of resources. But it may be that a third cause of death stressed the cave bear populations, since, after spreading throughout Europe during the middle Pleistocene, the early humans evolved into their own late Pleistocene robust variant, the Neanderthals, who probably competed with the cave bears for the occupation of the caves.

THE LAST ACT: THE SPREAD OF MODERN HUMANS

The first remains of an adult Neanderthal were discovered in 1848 in Gibraltar, although they were not described until 1864. In fact, it was not until 1856, when a less complete braincase was found in the Neander valley, in Germany, that paleontologists became aware of the significance of these findings. At first, the partial skull of Neanderthal started an argument, soon after the publication in 1859 of the first edition of *On the Origin of Species* by Charles Darwin. In his book *Man's Place in Nature* (published before Darwin's *The Descent of Man*), Thomas Huxley discussed the evolutionary implications of the Neanderthal findings in order to understand human origins.

From the very beginning, the evolutionary position of the Neanderthals was (and is still) a matter of intense debate, because of their puzzling mixture of archaic and advanced characteristics (figure 7.16). The Neanderthals shared with their immediate ancestor, *H. heidelbergensis,* a number of archaic, *erectus*-like characteristics, such as an elongated skull, with a low forehead and large superciliary arches. However, the shape of the braincase was different, not flat and low, as in the middle Pleistocene hominids, but rounded and high vaulted. The orbits were not quadrangular, as in other human species, but large and characteristically rounded. Although highly developed, the superciliary arches over the orbits were also different. They no longer formed a kind of visor but covered only the eyes and were interrupted at the level of the nose. The nose itself was prominent and wide, and very high, which has been interpreted as an adaptation allowing the isolation of the nasal cavities from the cold air of the late Pleistocene. The face was large and prominent, outward directed ("prognathism"). This was perhaps a consequence of the robust anterior dentition, especially in the incisors, which were prominent and adapted to chewing and preparing tough food before ingestion (most of the Neanderthals' incisors show characteristic wear not found in other human species). The shape of the Neanderthal mandible was also highly prognathous. The "advanced" position of the dentition in the mandible led to a unique feature of the Neanderthals: a wide space between the last lower third molar and the ascending ramus of the mandible (if feasible, a "fourth" molar could have been located in this region).

FIGURE 7.16 *Reconstructed life appearance of a male Neanderthal*

Compared with an average modern human, Neanderthals would have looked exceptionally robust and muscular. Muscle insertion areas in the limb bones were strongly developed, and the bones themselves were thick. Both sexes had a wide pelvis, with laterally flaring illia that indicate a wide torso and large body mass. Compared with modern humans, but also with their likely ancestor, *Homo heidelbergensis,* Neanderthals had rather short limbs, which was probably part of their adaptation to life in cold climates. Reconstructed height of male: 1.65 m.

The robustness of the Neanderthals was also evident in the body, which displayed strong, relatively short arms and legs. This was clearly an adaptation to the cold conditions of the steppe environments where the Neanderthals settled in the late Pleistocene. However, in the early twentieth century this anatomy was originally interpreted as a proof of the species's primitiveness. In a monograph, the French paleontologist Marcellin Boule exaggerated these characteristics to present Neanderthals as brutes, much different from modern humans. He described Neanderthals as primitive humans with short body proportions and imperfect bipedism. However, against this idea of primitiveness was the fact, already recognized by the first authors who analyzed the Neanderthal braincase, that the Neanderthals' cranial capacity was certainly larger than that of modern humans. Today we know that the mean encephalic volume of Neanderthal males was about 1,600 cc, while the female value was about 1,300 cc (against 1,500 and 1,200 cc, respectively, in modern humans). Moreover, the Neanderthals made elaborated stone tools known as the Mousterian Industry, which were certainly much more sophisticated than those of their predecessors (who, nevertheless, had developed the archaic Acheulian Industry). In cultural terms, the last Neanderthals were, in fact, quite close to the first modern humans, and we can find among them the first evidence of burials. Although the highest proportion of Neanderthal remains is concentrated in western and central Europe (for example, La Chappelle aux Saints and La Ferrasie, France; Monte Circeo, Italy; and Krapina, Croatia), this species covered a wide range, from southern Spain (Zafarraya) to the Levant (Tabun) and western Asia (Shanidar, Iran).

However, around 40,000 years ago, a new human type suddenly invaded Europe, completely replacing (or displacing?) the Neanderthals shortly afterward (around 30,000 years ago). These new humans were our own ancestors, often known as "modern humans." The anatomy of this species was very different from that of the Neanderthals, with tall bodies (about 1.80 m in the case of the first late Pleistocene European representatives, known as Cro-Magnon humans) and long, slender limbs. The cranial vault was high and characteristically rounded, much different from the elongated braincases seen in the Neanderthals and the *erectus*-like variants. The front was high, and the orbits were low and quadrangular, lacking the thick superciliary arches present in earlier human forms. The face was low and flat and not prognathous, as in the Neanderthals. The mandible presented a chin, which is a unique feature among the hominid species.

The first evidence of this new human type, found in different parts of Africa, date from between 200,000 and 100,000 years ago. A number of partial skulls, such as those of Ngaloba in Laetoli, Tanzania, and Djebel Irhoud, Morocco, although retaining some archaic features (such as superciliary arches), already displayed a number of character-istics that were in the direction of modern humans (rounded vault with a wide and flat face, among others). By around 100,000 years ago, the modern combination of features—short, rounded vault; smaller super-ciliary arches; prominent chin; and slender, tall anatomy—was fully present in a number of African findings. This is the case, for instance, of the skull of Klasies River, South Africa, and the partial skeleton of Omo Kibish I, Ethiopia. All these findings attest to the presence of mod-ern humans in Africa before 100,000 years ago. The tall anatomy of the first modern humans can be best explained as an adaptation to life in the savanna and the open landscapes of Africa, in clear contrast with the robust constitution of the cold-adapted Neanderthals.

However, the mixture of archaic and advanced characteristics in these early modern humans led to misleading conclusions about their origins, at a time when the dating of these findings was confusing. When in the 1920s and 1930s the first evidence of this early modern human type appeared in a number of caves in El Taboun and Mugharet, at Mount Carmel in Palestine, they were interpreted as intermediate forms between the Neanderthals and the European Cro-Magnons. The com-plete skulls of Skhul-V and Qafzeh showed an overall modern mor-phology but retained, as in the African "proto-moderns," some archaic features like moderately developed superciliary arches. The finding of true Neanderthals in other sites of the region, such as Tabun I, sup-ported this intermediate interpretation. Since at that time there was no secure method for the dating of these remains, it was assumed that the Neanderthals of Tabun I were, in fact, older than and preceded the early moderns in the occupation of the area, evolving later in situ. How-ever, when accurate dating by electronic spin resonance and other methods became available, the overall picture changed completely. The modern skulls of Qafzeh and Skhul are, in fact, older than most of the Neanderthals of the region, reaching an age of 120,000 to 100,000 years, close to that of the first modern humans of Africa. But the Ne-anderthal skulls of Kebara and Amud are younger, about 50,000 to 40,000 years old. The Neanderthal of Tabun, in turn, is older, reaching an age close to that of the remains of Skhul and Qafzeh. The rather simple original succession Neanderthal–modern human was replaced by a much more complicated one: modern human–Neanderthal, or Neanderthal–modern human–Neanderthal.

The Middle East succession seems to reflect more an ecological turn-over than a phylogenetic sequence. The supposed Neanderthal features of the remains at Skhul and Qafzeh are, in fact, archaic features inherited from the common ancestor of both human types: *H. ergaster, H. antecessor,* or *H. heidelbergensis.* Speaking in cladistic terms, they are primitive, ple-siomorphic characteristics that cannot be used for phylogenetic purposes. Neanderthals and modern humans clearly correspond to two different ecotypes: a robust one, well adapted to the cold conditions of glaciated Europe; and a tall, slender one, adapted to the open landscapes and savannas of Africa. In this context, the Middle East appears as a boundary area reflecting the climatic fluctuations of the late Pleistocene. The modern human–Neanderthal and Neanderthal–modern human–Neanderthal sequences recorded there reflect the latitudinal displacements of the boundary between the cold steppe to the north and the subdesert-savanna conditions to the south. Neanderthals and modern humans would have followed their own biotopes to the south or to the north, depending on the latitudinal fluctuations of the boundary between these two biomes.

Modern humans replaced Neanderthals about 30,000 years ago, before the set of events that led to the extinction of the late Pleistocene megafauna at the beginning of the present interglacial period. What caused such a replacement? Apparently, the more sophisticated culture of modern humans emerges as a plausible factor. The lithic artifacts of modern humans, included in the Aurignacian Industry, were technically more advanced than the Mousterian tools of the Neanderthals. And modern humans were clearly capable of symbolic expression, as reflected by the paintings they executed in caves in southern France and northern Spain. They fashioned not only tools but also small statues, necklaces, and other objects not directly useful for survival. They also buried their dead, accompanied by several of these objects, in sophisticated burials. Thus they appeared to be much more advanced than their Neanderthal cousins.

However, the evidence found with the last Neanderthal populations in France contradicts this interpretation. The finds at Saint Césaire, dated at around 36,000 years ago, yielded rather complete remains of a Neanderthal associated with an elaborated industry resembling the Aurignacian. This industry, called Chatelperronian, is more advanced than the Mousterian and indicates that at the end of their history the Neanderthals achieved a high technical level in the making of tools, similar to that of the first modern humans who entered Europe. Also, there are some indications that these last Neanderthals could have buried their dead as well, based on the evidence found in some classical Neanderthal sites such as Kebara and Taboun, Israel, and Shanidar, Iran. Therefore, the argument that modern humans replaced Neanderthals because of their higher cultural capabilities has to be questioned. But if culture cannot explain the Neanderthal–modern human succession, the replacement of one by the other at around 30,000 years ago can be explained only in terms of biological fitness. Since Neanderthals were the species established on the steppes of Europe for more than 80,000 years, the spread of modern humans was probably related to some kind of demographic revolution that enabled them to expand their range to the north of the Caucasus, over the whole of Europe. In other words, modern humans probably replaced Neanderthals because they were able to reproduce faster and in higher numbers than their robust, cold-adapted neighbors.

In contrast with the Neanderthals, modern humans not only settled in Europe but expanded their range over the whole of Eurasia and reached Australia about 40,000 years ago. Finally, at the end of the Pleistocene, they crossed the Bering Strait, which at that time was a wide dry region, and colonized the Americas, reaching territories never previously inhabited by any human species.

THE LAST MASS EXTINCTION

At the end of the last glacial period, 14,000 years ago, the ice sheets started to retreat once again, and 10,000 years ago a new interglacial phase, the Holocene, began. The temperatures suddenly increased between 5 and 7°C, and the sea rose about 120 m to its present level. In Eurasia, the steppe-tundra, the original habitat occupied by mammoths and woolly rhinos, split and became decimated. An uninterrupted belt of coniferous trees, the taiga, extended throughout Europe and eastern Asia. To the north, the narrow belt of the Arctic tundra was suitable for only a few steppe-tundra species, such as reindeer and lemmings. To the south, the dry steppe of Central Asia was home to a number of medium-size ungulates, such as horses, bison, and antelopes, as well as several rodent species. Others, finally, survived in the Arctic regions of North America (musk ox). But the largest representatives of the late Pleistocene megafauna became completely extinct. The last mammoths in Europe are recorded in Denmark and Sweden about 13,000 years ago. In Siberia, the last record of these proboscideans is dated at 10,000 years ago, although a number of them probably persisted until less than 4,000 years ago on some small islands of the Arctic Ocean north of Siberia. For the mammoth and the woolly rhino, the reduction of their natural habitat may have led to a decrease in their population, thus enhancing the probability of extinction.

However, an alternative cause may have played a main role in the late Pleistocene megafaunal extinction—that is, the spread of modern humans over the whole planet. The expansion of modern humans seems to have been fatal not only for Neanderthals but also for other significant large ice-age mammals. Although there is no direct evidence that humans hunted the cave bear, this species disappeared in Europe around 18,000 years ago, at the peak of the last glacial phase and well before the advent of the Holocene interglacial period. Even earlier, around 40,000 years ago, modern humans reached Australia from Asia.

This continent contained at that time a rich marsupial fauna that was the product of millions of years of isolated evolution since the early Cenozoic. However, the largest species in Australia became extinct soon after the entry of humans. More impressive is the evidence from North America. On this continent, the melting of the glaciers opened a new ice-free corridor with eastern Asia and enabled the entry of the first paleo-Indians around 14,000 years ago. Although North America underwent less dramatic changes than Eurasia during the Pleistocene, in a short time, between 11,000 and 10,000 years ago, about thirty genera of large mammals became extinct, including mastodons, mammoths, tapirs, horses, artiodactyls, and edentates of South American origin (as well as the cohort of carnivores associated with these prey). In Africa, the oldest continent inhabited by modern humans, a similar set of extinctions of large herbivores and carnivores began as early as the middle Pleistocene. *Homo sapiens* appears unique because of its ability to exterminate other species. This capability does not seem characteristic of other hominid species. Neanderthals largely coexisted with several megafaunal species without causing their extinction. And in Africa, advanced hominids like *H. ergaster* coexisted with several large mammalian genera of Miocene or Pliocene origin until their demise during the middle Pleistocene.

MAMMALS BETWEEN TWO MASS EXTINCTIONS

Later in the Holocene, the extinction wave has continued. After modern humans settled the Mediterranean islands, the endemic faunas developed there during the Pleistocene rapidly became extinct. A number of late Pleistocene survivors, such as the lion and the leopard, also disappeared from Europe in historical times. Others were severely endangered, having become considerably restricted in their original habitat, as was the case with the brown bear, the lynx, and the bison. The restriction of these species was the effect not only of modern humans but also of the mammals associated with them, since *H. sapiens* has become an agent not only of extinction but also of evolution. With the advent of the Neolithic revolution, some mammalian species were favored by humans, who used them for their benefit. They were modified by active selection of the varieties and morphotypes most useful for human purposes. This is how the domestic varieties of dogs, cats, cattle, sheep, pigs, and goats arose. This was a kind of symbiosis, since these species also benefited from an expanded population range. Other species, although not desired or domesticated, also benefited from human expansion, broadening their distribution in historical times. This was the case with mice (*Mus musculus*) and rats (*Rattus rattus* and *R. norvergicus*), the latter having entered Europe during the Middle Ages and becoming the agent of the terrible plagues that decimated the European population in the fourteenth century.

Finally, humans domesticated themselves. After thousands of years of dispersal and cultural differentiation, followed by selective inter-breeding, the present diversity of human variants can be regarded as a peculiar case of human selection. This is a genetic treasure that has resulted in an extraordinary variety of morphotypes and that has survived thousands of years of mass death and systematic destruction during wars. Therefore, even with the advent of modern humans, evolution goes on.

But the main effect of human activities is still to come. With the beginning of the Industrial Revolution in the nineteenth century, modern humans began ejecting into the atmosphere large amounts of CO_2 and other greenhouse gases that had remained stored in oceanic and continental reservoirs for millions of years. The main effect of the emission of these greenhouse gases is a progressive warming of the planet, which will probably lead to the interruption of the glacial–interglacial dynamic started in the late Pliocene. As a consequence, the Earth system is going to return to much warmer climatic conditions. The melting of the Arctic and western Antarctic Ice Sheets is a feasible consequence of this warming, as well as a general rise in the global sea level.

But certainly, this is not a new scenario for the planet. For instance, we know that a sudden and intense warming took place at the end of the Paleocene, when the early Eocene tropical forest expanded over large regions of the Earth. What will be the condition of the biosphere during the next thousands of years? Will our future climate resemble that of the middle Pleistocene mild interglacial periods or, better, those of the early Pliocene climatic optimum? Are we going to face the even warmer conditions of the early Eocene, when no ice sheet covered Antarctica? Perhaps it is too early to answer such questions.

Meanwhile, evolution must go on. Mammals are now facing their second mass extinction. When the first took place 65 million years ago, they were represented by small, generalist forms capable of surviving long periods of stress. Now the situation is different, comparable to that of the last dinosaurs. In the new conditions imposed by humans, our decimated mammalian faunas can hardly survive without our help, and our behavior should change in the future to preserve these biological treasures. Otherwise, they will pass also into the fossil record and become the subject of a future version of this book.

BIBLIOGRAPHY

This bibliography does not pretend to be exhaustive and merely refers to those works whose information has been particularly useful in the writing of this book. The works marked with an asterisk (*) have been used in inferring the body weight and lifestyle of the group involved.

GENERAL WORKS

Agustí, J., and S. Moyà-Solà. 1990. "Mammal extinctions in the Vallesian (Upper Miocene)." *Lecture Notes in Earth Science* 30:425–32.

Agustí, J., L. Rook, and P. Andrews, eds. 1999. *Evolution of Neogene Terrestrial Ecosystems in Europe*. Cambridge: Cambridge University Press.

*Alcalá, L. 1994. *Macromamíferos neógenos de la fosa de Alfambra-Teruel*. Zaragoza: Instituto de Estudios Riojanos.

Andersen, B. G., and H. W. Borns, Jr. 1994. *The Ice Age World*. Oslo: Scandinavian University Press.

Azzaroli, A., C. De Giuli, G. Ficcarelli, and D. Torre. 1988. "Late Pliocene to early mid-Pleistocene mammals in Eurasia: Faunal succession and dispersal events." *Palaeogeography, Palaeoclimatology, Palaeoecology* 66:77–100.

Barry, J. C., N. M. Johnson, S. Mahmood-Raza, and L. L. Jacobs. 1985. "Neogene mammalian faunal change in southern Asia: Correlations with climatic, tectonic, and eustatic events." *Geology* 13:637–40.

Bernor, R., V. Fahlbusch, and W. Mittman, eds. 1996. *The Evolution of Western Eurasian Neogene Mammal Faunas*. New York: Columbia University Press.

Brunet, M. 1979. *Les Grands Mammifères chefs de file de l'immigration oligocène et le problème de la limite Eocène–Oligocène en Europe*. Paris: Editions Fondation Singer-Polignac.

Crusafont, M. 1950. "La cuestión del llamado Meótico español." *Arrahona* 1: 3–9.

*Damuth, J., and B. MacFadden, eds. 1990. *Body Size in Mammalian Paleobiology*. Cambridge: Cambridge University Press.

Franzen, J. L., and G. Storch. 1999. "Late Miocene mammals from central Europe." In J. Agustí, L. Rook, and P. Andrews, eds., *Evolution of Neogene Terrestrial Ecosystems in Europe*. Cambridge: Cambridge University Press.

Godinot, M. 1982. "Aspects nouveaux des échanges entre les faunes mammaliennes d'Europe et d'Amérique du Nord à la base de l'Eocène." *Geobios* 6:403–12.

Guerin, C., and M. Patou-Mathis, eds. 1996. *Les Grands Mammifères Plio-Pléistocènes d'Europe*. Paris: Masson.

Hooker, J. J. 1989. "British mammals in the Tertiary period." *Biological Journal of the Linnean Society* 38:9–21.

Janis, C. M., and J. Damuth. 1991. "Mammals." In K. J. McNamara, ed., *Evolutionary Trends*. London: Belhaven Press.

Kahlke, R.-D. 1999. *The History of the Origin, Evolution, and Dispersal of the Late Pleistocene* Mammuthus-Coelodonta *Faunal Complex in Eurasia (Large Mammals)*. Rapid City, S.D.: Fenske.

Nadachowski, A., and L. Wenderlin, eds. 1996. *Neogene and Quaternary Mammals of the Palaearctic*. Acta Zoologica Cracowiensa, no. 39. Krakow: Institute of Systematics and Evolution of Animals.

Pilbeam, D., M. Morgan, J. C. Barry, and L. J. Flynn. 1996. "European MN units and the Siwalik faunal sequence of Pakistan." In R. Bernor, V. Fahlbusch, and W. Mittman, eds., *The Evolution of Western Eurasian Neogene Mammal Faunas*. New York: Columbia University Press.

Prothero, D. R. 1994. *The Eocene–Oligocene Transition*. New York: Columbia University Press.

Rössner, G. E., and K. Heissig, eds. 1999. *The Miocene Land Mammals of Europe*. Munich: Pfeil.

Russell, D. E. 1964. "Les Mammifères paléocènes d'Europe." *Mémoires du Musée National d'Histoire Naturelle* 12:1–324.

Russell, D. E., J. L. Hartenberger, C. Pomerol, S. Sen, N. Schmidt-Kittler, and M. Vianey-Liaud. 1982. "Mammals and stratigraphy: The Paleogene of Europe." *Palaeovertebrata* 1–77.

Schaal, S., and W. Ziegler, eds. 1992. *Messel: An Insight into the History of Life and of the Earth*. Oxford: Clarendon Press.

Schmidt-Kittler, N., ed. "International Symposium on Mammalian Biostratigraphy and Palaeoecology of the European Paleogene." *Müchner Geowissencheften Abhandlungen* 10:1–311.

Tchernov, E. L. Ginsburg, P. Tassy, and N. F. Goldsmith. 1987. "Miocene mammals from the Negev (Israel)." *Journal of Vertebrate Paleontology* 7:284–310.

PALEOGEOGRAPHY AND CLIMATE EVOLUTION

Agustí, J., and L. Wenderlin, eds. 1995. *Influence of Climate on Faunal Evolution in the Quaternary*. Acta Zoologica Cracowiensa, no. 38. Krakow: Institute of Systematics and Evolution of Animals.

Antunes, M. T., and J. Pais. 1984. "Climate during the Miocene in Portugal and its evolution." *Paléobiologie continentale* 14:75–90.

Axelrod, D. I. 1975. "Evolution and biogeography of Madrean-Tethyan sclerophyll vegetation." *Annals of the Missouri Botanical Garden* 62:280–334.

Axelrod, D. I., and P. H. Raven. 1978. "Late Cretaceous and Tertiary vegetation history of Africa." In M. J. A. Werger, ed., *Biogeography and Ecology of Southern Africa.* The Hague: Junk.

Bessedik, M. 1984. "The early Aquitanian and Upper Langhian–Lower Serravallian environments in the north-western Mediterranean region." *Paléobiologie continentale* 14:153–80.

Bessedik, M., and L. Cabrera. 1985. "Le Couple récif-mangrove à Sant Pau d'Ordal (Vallès-Penedés, Espagne), témoin du maximum transgressif en Méditérranée nord occidentale (Burdigalien supérieur-Langhien inférieur)." *Newsletter of Stratigraphy* 14:20–35.

Bonnefille, R. 1995. "A reassessment of the Plio-Pleistocene pollen Record of East Africa." In E. S. Vrba, G. H. Denton, and T. C. Partridge, eds., *Paleoclimate and Evolution, with Emphasis on Human Origins.* New Haven, Conn.: Yale University Press.

Cerling, T. E. 1992. "Development of grasslands and savannas in East Africa during the Neogene." *Palaeogeography, Palaeoclimatology, Palaeoecology* 97: 241–47.

Cerling, T. E., J. R. Harris, B. J. MacFadden, M. G. Leakey, J. Quade, V. Eisenmann, and J. R. Ehleringer. 1997. "Global vegetation change through the Miocene/Pliocene boundary." *Nature* 389:153–58.

Collinson, M. E., and J. Hooker. 1987. "Vegetational and mammalian faunal changes in the early Tertiary of southern England." In E. M. Friis, W. G. Chaloner, and P. R. Crane, eds., *The Origins of Angiosperms and Their Biological Consequences.* Cambridge: Cambridge University Press.

Dupont, L., and S. Leroy. 1995. "Steps toward drier climatic conditions in north-western Africa during the Upper Pliocene." In E. S. Vrba, G. H. Denton, and T. C. Partridge, eds., *Paleoclimate and Evolution, with Emphasis on Human Origins.* New Haven, Conn.: Yale University Press.

Flower, B. P., and J. P. Kennett. 1994. "The middle Miocene climatic transition: East Antarctic ice sheet development, deep ocean circulation and global carbon cycling." *Palaeogeography, Palaeoclimatology, Palaeoecology* 108:537–55.

Fricke, H. C., W. C. Clyde, J. R. O'Neil, and P. D. Gingerich. 1998. "Evidence for rapid climate change in North America during the latest Paleocene thermal maximum: Oxygen isotope compositions of biogenic phosphate from the Bighorn Basin (Wyoming)." *Earth Planetary Science Letters* 160:193–208.

Gladenkov, Y. B. 1992. "North Pacific: Neogene biotic and abiotic events." In R. Rsuchi and J. C. Ingle, Jr., eds., *Pacific Neogene Environmental Evolution and Events.* Tokyo: University of Tokyo Press.

Goldsmith, N. F., F. Hisrch, G. M. Friedman, E. Tchernov, B. Derin, E. Gerry, A. Horowitz, and G. Weinberger. 1988. "Rotem mammals and Yeroham crassostreids: Stratigraphy of the Hazeva Formation (Israel) and the paleogeography of Miocene Africa." *Newsletter of Stratigraphy* 20:73–90.

Haq, B. U., J. Hardenbol, and P. R. Vail. 1987. "Chronology of fluctuating sea levels since the Triassic." *Science* 235:1156–67.

Harris, J. 1993. "Ecosystem structure and growth of the African savanna." *Global and Planetary Change* 8:231–48.

Hornibrook, N. D. B. 1992. "New Zealand Cenozoic marine paleoclimates: A review based on the distribution of some shallow water and terrestrial biota." In R. Rsuchi and J. C. Ingle, Jr., eds., *Pacific Neogene Environmental Evolution and Events.* Tokyo: University of Tokyo Press.

Horowitz, A. 1989. "Continuous pollen diagrams for the last 3.5 m.y. from Israel: Vegetation, climate and correlation with the oxygen isotope record." *Palaeogeography, Palaeoclimatology, Palaeoecology* 72:63–78.

Jones, R. Wynn. 1999. "Marine invertebrate (chiefly foraminiferal) evidence for the palaeogeography of the Oligocene–Miocene of western Eurasia, and consequences for terrestrial vertebrate migration." In J. Agustí, L. Rook, and P. Andrews, eds., *Evolution of Neogene Terrestrial Ecosystems in Europe.* Cambridge: Cambridge University Press.

Keller, G., and J. A. Barron. 1983. "Paleoceanographic implications of Miocene deep-sea hiatuses." *Geological Society of America Bulletin* 94:590–613.

Miller, K. G., R. G. Fairbanks, and G. S. Mountain. 1987. "Tertiary oxygen synthesis, sea level history, and continental margin erosion." *Paleoceanography* 2:1–19.

Miller, K. G., J. D. Wright, and R. G. Fairbanks. 1991. "Unlocking the icehouse: Oligocene–Miocene oxygen isotopes, eustacy and margin erosion." *Journal of Geophysical Research* 96:6829–48.

Morgan, M. E., J. D. Kingston, and B. D. Marino. 1994. "Carbon isotopic evidence for the emergence of C4 plants in the Neogene from Pakistan and Kenya." *Nature* 367:162–65.

Nichols, D. J., D. M. Jarzen, C. J. Orth, and P. Q. Oliver. 1986. "Palynological and iridium anomalies at Cretaceous–Tertiary boundary, south-central Saskatchewan." *Science* 231:714–17.

Quade, J., T. E. Cerling, and J. R. Bowman. 1989. "Development of Asian monsoon revealed by marked ecological shift during the latest Miocene in northern Pakistan." *Nature* 342:163–64.

Roberts, N. 1984. "Pleistocene environments in time and space." In R. Foley, ed., *Hominid Evolution and Communnity Ecology.* San Diego: Academic Press.

Rögl, F. 1999. "Mediterranean and Paratethys palaeogeography during the Oligocene and Miocene." In J. Agustí, L. Rook, and P. Andrews, eds., *Evolution of Neogene Terrestrial Ecosystems in Europe.* Cambridge: Cambridge University Press.

Sanz de Siria, A. 1997. "La macroflora del vallesiense superior de Terrassa (Barcelona)." *Paleontologia i evolució* 30–31:247–68.

Shackleton, N. J. 1995. "New data on the evolution of Pliocene climatic variability." In E. S. Vrba, G. H. Denton, and T. C. Partridge, eds., *Paleoclimate and Evolution, with Emphasis on Human Origins.* New Haven, Conn.: Yale University Press.

Shackleton, N. J., and J. P. Kennett. 1975. "Late Cenozoic oxygen and carbon isotope changes at DSDP site 284: Implications for glacial history of the Northern Hemisphere and Antarctica." In J. P. Kennett, R. E. Houtz, P. B. Andrews, A. R. Edwards, V. A. Gostin, M. Hajos, M. A. Hampton, D. G. Jenkins, S. V. Margolis, A. T. Ovenshine, and K. Perch-Nielsen, eds., *Initial Reports of the Deep Sea Drilling Project.* Washington, D.C.: Government Printing Office.

Solounias, N., J. M. Plavcan, J. Quade, and L. Witmer. 1999. "The paleoecology of the Pikermian Biome and the savanna myth." In J. Agustí, L. Rook, and P. Andrews, eds., *Evolution of Neogene Terrestrial Ecosystems in Europe.* Cambridge: Cambridge University Press.

Suc, J. P., F. Diniz, S. Leroy, C. Poumont, A. Bertini, L. Dupont, M. Clet, E. Bessais, Z. Zheng, S. Fauquette, and J. Ferrier. 1995. "Zanclean (Brunssumian) to early Piazencian (early-middle Reuverian) climate from 4§ to 54§ north latitude (West Africa, West Europe and west Mediterranean areas)." *Mededelingen Rijks geologische Dienst* 52:43–56.

Suc, J. P., S. Fauquette, M. Bessedik, A. Bertini, Z. Zheng, G. Clauzon, D. Suballyova, F. Diniz, P. Quézel, N. Feddi, M. Clet, E. Bessais, N. Bachiri, H. Meon, and N. Comborieu-Nebout. 1999. "Neogene vegetation changes in West European and west circum-Mediterranean areas." In J. Agustí, L. Rook, and P. Andrews, eds., *Evolution of Neogene Terrestrial Ecosystems in Europe.* Cambridge: Cambridge University Press.

Vincent, E., and W. H. Berger. 1985. "Carbon dioxide and polar cooling in the Miocene: The Monterey Hypothesis." In E. T. Sundquist and W. S. Broecker, eds., "The Carbon Cycle and Atmospheric CO_2: Natural Variations Archean to Present." *Geophysical Monographs* 32:455–68.

Volkova, V. S., I. A. Kul'kova, and A. F. Fradkina. 1986. "Palynostratigraphy of non-marine Neogene in North Asia." *Review of Palaeobotany and Palynology* 48:415–24.

Woodruf, F. 1985. "Changes in Miocene deep-sea benthic foraminiferal distribution in the Pacific Ocean: Relationship to paleoceanography." *Geological Society of America Memoirs* 163:131–76.

Woodruf, F., and S. Savin. 1989. "Miocene deepwater oceanography." *Paleoceanography* 4:87–140.

Wrenn, J. H., J. P. Suc, and S. A. G. Leroy, eds. *The Pliocene: Time of Change.* Dallas: American Association of Stratigraphic Palynologists Foundation.

Wright, J. D., K. G. Miller, and R. G. Fairbanks. 1992. "Early and middle Miocene stable isotopes: Implications for deepwater circulation and climate." *Paleoceanography* 7:357–89.

Zubakov, V. A., and I. I. Borzenkova. 1990. *Global Paleoclimate of the Late Cenozoic.* Amsterdam: Elsevier.

MULTITUBERCULATES, ARCHAIC THERIANS, INSECTIVORES, AND BATS

Butler, P. M. 1980. "The giant erinaceid insectivore *Deinogaleryx* Freudenthal, from the Upper Miocene of Gargano, Italy." *Scripta Geologica* 57:1–72.

Butler, P. M. 1988. "Phylogeny of the Insectivora." In M. J. Benton, ed. *The Phylogeny and Classification of the Tetrapods.* Vol. 2, *Mammals.* Oxford: Clarendon Press.

Habersetzer, J., G. Richter, and G. Storch. 1992. "Bats: Already highly specialized insect predators." In S. Schaal and W. Ziegler, eds., *Messel: An Insight into the History of Life and of the Earth.* Oxford: Clarendon Press.

Koenigswald, W. von, and G. Storch. 1992. "The marsupials: Inconspicuous opossums." In S. Schaal and W. Ziegler, eds., *Messel: An Insight into the History of Life and of the Earth.* Oxford: Clarendon Press.

Koenigswald, W. von, G. Storch, and G. Richter. 1992. "Primitive insectivores, extraordinary hedgehogs and long fingers." In S. Schaal and W. Ziegler, eds., *Messel: An Insight into the History of Life and of the Earth.* Oxford: Clarendon Press.

Pélaez-Campomanes, P., N. López-Martínez, M. A. Alvarez-Sierra, and R. Daams. 2000. "The earliest mammal of the European Paleocene: The multituberculate *Hainina.*" *Journal of Paleontology* 74:701-11.

Radulesco, C., and P. Samson. 1986. "Précisions sur les affinités des multituberculés (Mammalia) du Cretacé supérieur de Roumanie." *Comptes rendus de l'Académie de Sciences de Paris,* 2nd ser., 303:1825–30.

Reumer, J. W. F. 1999. "Shrews (Mammalia, Insectivora, Soricidae) as paleoclimatic indicators in the European Neogene." In J. Agustí, L. Rook, and P. Andrews, eds., *Evolution of Neogene Terrestrial Ecosystems in Europe.* Cambridge: Cambridge University Press.

Russell, D. E., and M. Godinot. 1988. "The Paroxyclaenidae (Mammalia) and a new form from the early Eocene of Palette, France." *Paläontologische Zeitschrift* 62:319–31.

Storch, G. 1996. "Paleobiology of Messel Erinaceomorphs." *Palaeovertebrata* 25:215–24.

Storch, G. 1999. "Order Chiroptera." In G. E. Rössner and K. Heissig, eds., *The Miocene Land Mammals of Europe.* Munich: Pfeil.

Wójcik, J. M., and M. Wolsan, eds. 1998. *Evolution of Shrews.* Bialowieza: Mammal Research Institute, Polish Academy of Sciences.

Ziegler, R. 1999a. "Order Insectivora." In G. E. Rössner and K. Heissig, eds., *The Miocene Land Mammals of Europe.* Munich: Pfeil.

Ziegler, R. 1999b. "Order Marsupialia: *Amphiperatherium,* the last European Opossum." In G. E. Rössner and K. Heissig, eds., *The Miocene Land Mammals of Europe.* Munich: Pfeil.

PRIMATES

Alpagut, B., P. Andrews, M. Fortelius, J. Kappelman, I. Temizsoy, H. Celebi, and E. Lindsay. 1996. "A new specimen of *Ankarapithecus meteai* from the Sinap Formation of central Anatolia." *Nature* 382:349–51.

*Andrews, P., T. Harrison, E. Delson, R. Bernor, and L. Martin. 1996. "Distribution and biochronology of European and Southwest Asian Miocene catarrhines." In R. Bernor, V. Fahlbusch, and W. Mittman, eds., *The Evolution of Western Eurasian Neogene Mammal Faunas.* New York: Columbia University Press.

Bermúdez de Castro, J. M., J. L. Arsuaga, E. Carbonell, A. Rosas, I. Martínez, and M. Mosquera. 1997. "A hominid from the Lower Pleistocene of Atapuerca, Spain: Possible ancestor to Neandertals and modern humans." *Science* 276:1392–95.

Bonis, L. de. 1995. "Primates hominoïdes du Miocène supérieur d'Europe." *Ethnologia,* n.s., 5:15–28.

Bonis, L. de, G. Bouvrain, D. Geraards, and G. Koufos. 1990. "New hominid skull material from late Miocene of Macedonia in northern Greece." *Nature* 345:712–14.

*Feagle, J. C. 1999. *Primate Adaptation and Evolution.* San Diego: Academic Press.

Godinot, M. 1998. "A summary of adapiform systematics and phylogeny." *Folia Primatologica* 69(suppl. 1):218–49.

Godinot, M., and M. Mahboubi. 1992. "Earliest known simian primate found in Algeria." *Nature* 357:324–26.

Köhler, M., and S. Moyà-Solà. 1997. "Ape-like or hominid-like? The positional behavior of *Oreopithecus bambolii* reconsidered." *Proceedings of the National Academy of Sciences USA* 94:11747–50.

Köhler, M., S. Moyà-Solà, and D. M. Alba. 2000. "*Macaca* (Primates, Cercopithecidae) from the late Miocene of Spain." *Journal of Human Evolution* 38:447–52.

Lordkipanidze, D., O. Bar-Yosef, and M. Otte, eds. 2000. *Early Humans at the Gates of Europe.* Liège: Etudes et Recherches Archéologiques de l'Université de Liège.

*Moyà, S., and M. Kohler. 1996. "A *Dryopithecus* skeleton and the origin of great-ape locomotion." *Nature* 379:156–59.

Tattersall, I., E. Delson, and J. van Couvering, eds. 1988. *Encyclopedia of Human Evolution and Prehistory.* New York: Garland.

RODENTS AND LAGOMORPHS

Agustí, J. 1990. "The Miocene rodent succession in eastern Spain: A zoogeographical appraisal." In E. Lindsay, V. Fahlbusch, and P. Mein, eds. *European Neogene Mammal Chronology.* New York: Plenum.

Chaline, J., and P. Mein. 1979. *Les Rongeurs et l'évolution.* Paris: Doin.

Engesser, B. 1999. "Family Eomyidae." In G. E. Rössner and K. Heissig, eds., *The Miocene Land Mammals of Europe.* Munich: Pfeil.

Fejfar, O., and W.-D. Heinrich. 1990. *International Symposium of the Evolution, Phylogeny and Biostratigraphy of Arvicolids (Rodentia, Mammalia).* Prague: Czechoslovakian Geological Survey.

Hugueney, M. 1999. "Family Castoridae." In G. E. Rössner and K. Heissig, eds., *The Miocene Land Mammals of Europe.* Munich: Pfeil.

López, N., and L. Thaler. 1974. "Sur le plus ancien lagomorphe européen et la 'Grande Coupure' de Stehlin." *Palaeovertebrata* 6:243–51.

Luckett, W. P., and J. L. Hartenberger, eds. 1985. *Evolutionary Relationships Among Rodents.* New York: Plenum.

Martin, R. A., and A. Tesakov, eds. 1998. "The early evolution of *Microtus.*" *Paludicola* 2:1–129.

Nesin, V., and V. A. Topachevski. 1999. "The late Miocene small mammal succession in Ukraine." In J. Agustí, L. Rook, and P. Andrews, eds., *Evolution of Neogene Terrestrial Ecosystems in Europe.* Cambridge: Cambridge University Press.

Schmidt-Kittler, N., and M. Vianey-Liaud. 1979. "Evolution des Aplodontidae oligocènes européens." *Palaeovertebrata* 9:33–82.

Storch, G., B. Engesser, and M. Wuttke. 1996. "Oldest fossil record of gliding in rodents." *Nature* 379:439–41.

Vianey-Liaud, M. 1974. "*Palaeosciurus goti,* n. sp., écureil terrestre de l'Oligocène moyen du Quercy: Données nouvelles sur l'apparition des sciuridés en Europe." *Annales de paleontologie vertébrés* 60:103–22.

CREODONTS AND CARNIVORES

Alcalá, L., J. Morales, and D. Soria. 1990. "El registro fósil neógeno de los carnívoros (Creodonta y Carnívora, Mammalia) de España." *Paleontologia i evolució* 23:67–73.

Argant, A., and R. Ballesio. 1996. "Famille des Felidae." In C. Guerin and M. Patou-Mathis, eds., *Les Grands Mammifères Plio-Pléistocènes d'Europe*. Paris: Masson.

Argant, A., and E. Crégut. 1996. "Famille des Ursidae." In C. Guerin and M. Patou-Mathis, eds., *Les Grands Mammifères Plio-Pléistocènes d'Europe*. Paris: Masson.

Bonis, L. de, and E. Cirot. 1992. "Le Garouillas et les sites contemporains (Oligocene, MP 25) des phosphorites du Quercy (Lot, Tarn-et Garonne, France) et leurs faunes de vertébrés. 7. Carnivores." *Palaeontographica*, A 236:135–49.

Bonis, L. de, S. Peigné, and M. Hugueney. 1999. "Carnivores féloïdes de l'Oligocène supérieur de Coderet-Bransat (Allier, France)." *Bulletin de la Société Géologique de France* 170:939–49.

Cirot, E., and L. de Bonis. 1992. "Revision du genre *Amphicynodon*, carnivore de l'Oligocene." *Palaeontographica*, A 220:103–30.

Crégut, E. 1996a. "Famille des Canidae." In C. Guerin and M. Patou-Mathis, eds., *Les Grands Mammifères Plio-Pléistocènes d'Europe*. Paris: Masson.

Crégut, E. 1996b. "Famille des Hyaenidae." In C. Guerin and M. Patou-Mathis, eds., *Les Grands Mammifères Plio-Pléistocènes d'Europe*. Paris: Masson.

Geraards, D., and E. Guleç. 1997. "Relationships of *Barbourofelis piveteaui* (Ozansoy, 1965), a late Miocene nimravid (Carnivora, Mammalia) from central Turkey." *Journal of Vertebrate Paleontology* 17:370–75.

Ginsburg, L. 1999a. "Order Carnivora." In G. E. Rössner and K. Heissig, eds., *The Miocene Land Mammals of Europe*. Munich: Pfeil.

Ginsburg, L. 1999b. "Order Creodonta." In G. E. Rössner and K. Heissig, eds., *The Miocene Land Mammals of Europe*. Munich: Pfeil.

*Legendre, S., and C. Roth. 1988. "Correlation of carnassial tooth size and body weight in recent carnivores (Mammalia)." *Historical Biology* 1:85–98.

Peigné, S., and L. de Bonis. 1999a. "The genus *Stenoplesictis* Filhol (Mammalia, Carnivora) from the Oligocene deposits of the phosphorites of Quercy, France." *Journal of Vertebrate Paleontology* 19:566–75.

Peigné, S., and L. de Bonis. 1999b. "Le Premier Crâne de *Nimravus* (Mammalia, Carnivora) d'Eurasie et ses relations avec *N. brachyops* d'Amerique du Nord." *Revue paléobiologie de Genève* 18:57–67.

Springhorn, R. 1992. "Carnivores: Agile climbers and prey catchers." In S. Schaal and W. Ziegler, eds., *Messel: An Insight into the History of Life and of the Earth*. Oxford: Clarendon Press.

Turner, A., and M. Antón. 1996. "The giant hyaena *Pachycrocuta brevirostris* (Mammalia, Carnivora, Hyaenidae)." *Geobios* 29:455–68.

*Turner, A., and M. Antón. 1997. *The Big Cats and Their fossil relatives*. New York: Columbia University Press.

*Viranta, S. 1996. "European Miocene Amphicyonidae: Taxonomy, systematics and ecology." *Acta Zoologica Fennica* 204:1–61.

Wenderlin, L. 1996. "Carnivores, exclusive of Hyaenidae, from the later Miocene of Europe and western Asia." In R. Bernor, V. Fahlbusch, and W. Mittman, eds., *The Evolution of Western Eurasian Neogene Mammal Faunas.* New York: Columbia University Press.

*Wenderlin, L., and N. Solounias. 1991. "The Hyaenidae: taxonomy, systematics and evolution." *Fossils & Strata* 30:1–104.

Wenderlin, L., and N. Solounias. 1996. "The evolutionary history of hyaenas in Europe and western Asia during the Miocene." In R. Bernor, V. Fahlbusch, and W. Mittman, eds., *The Evolution of Western Eurasian Neogene Mammal Faunas.* New York: Columbia University Press.

PERISSODACTYLS

Bernor, R., and M. Armour-Chelu. 1999. "Family Equidae." In G. E. Rössner and K. Heissig, eds., *The Miocene Land Mammals of Europe.* Munich: Pfeil.

Bonis, L. de. 1995. "Le Garouillas et les sites contemporains (Oligocene, MP 25) des phosphorites du Quercy (Lot, Tarn-et Garonne, France) et leurs faunes de vertébrés. 11. Perissodactyla: Chalicotherioidea." *Palaeontographica,* A 236:191–204.

Bonis, L. de, and M. Brunet. 1995. "Le Garouillas et les sites contemporains (Oligocene, MP 25) des phosphorites du Quercy (Lot, Tarn-et Garonne, France) et leurs faunes de vertébrés. 10. Perissodactyla: Allaceropinae et Rhinocerotidae." *Palaeontographica,* A 236:177–90.

Coombs, M. C. 1989. "Interrelationships and diversity in the Chalicotheridae." In D. R. Prothero and R. M. Schoch, eds., *The Evolution of the Perissodactyls.* Oxford: Oxford University Press.

Eisenmann, V. 1991. "Les Chevaux quaternaires européens (Mammalia, Perissodactyla): Taille, typologie, biostratigraphie et taxonomie." *Geobios* 24: 747–759.

Eisenmann, V., and C. Guerin. 1992. "*Tapirus priscus* Kaup from the Upper Miocene of western Europe: Palaeontology, biostratigraphy and palaeoecology." In J. Agustí, ed., "Global Events and Neogene Evolution of the Mediterranean." *Paleontologia i evolució* 24–25:113–22.

Fortelius, M. 1990. "Rhinocerotidae from Paçalar, middle Miocene of Anatolia (Turkey)." *Journal of Human Evolution* 19:489–508.

Fortelius, M., P. Mazza, and B. Sala. 1993. "*Stephanorhinus* (Mammalia, Rhinocerotidae) of western European Pleistocene, with a revision of *S. etruscus* (Falconer, 1868)." *Paleontographia italica* 80:63–155.

Franzen, J. L. 1995. "Die Equoidea des europäischen Mitteleozäns (Geiseltalium)." *Hallesches Jahrbuch Geowisseuschaften,* B 17:31–45.

Froelich, D. J. 1999. "Phylogenetic systematics of basal perissodactyls." *Journal of Vertebrate Paleontology* 19:140–59.

Garcés, M., L. Cabrera, J. Agustí, and J. M. Parés. 1997. "Old World first appearence datum of 'Hipparion' horses: Late Miocene large mammal dispersal and global events." *Geology* 25:19–22.

*Heissig, K. 1989. "The Rhinocerotidae." In D. R. Prothero and R. M. Schoch, eds., *The Evolution of the Perissodactyls.* Oxford: Oxford University Press.

Heissig, K. 1999a. "Family Chalicotheridae." In G. E. Rössner and K. Heissig, eds., *The Miocene Land Mammals of Europe.* Munich: Pfeil.

Heissig, K. 1999b. "Family Rhinocerotidae. In G. E. Rössner and K. Heissig, eds., *The Miocene Land Mammals of Europe.* Munich: Pfeil.

Heissig, K. 1999c. "Family Tapiridae." In G. E. Rössner and K. Heissig, eds., *The Miocene Land Mammals of Europe.* Munich: Pfeil.

Hooker, J. J. 1994. "The beginning of the equoid radiation." In M. J. Benton and D. B. Norman, eds., "Vertebrate Palaeobiology." *Zoological Journal of the Linnean Society* 112:29–63.

Janis, C. M. 1976. "The evolutionary strategy of the Equidae and the origins of rumen and cecae digestion." *Evolution* 30:757–74.

MacFadden, B. J. 1992. *Fossil Horses.* Cambridge: Cambridge University Press.

Prothero, D. R., C. Guerin, and E. Manning. 1989. "The history of the Rhinocerotoidea." In D. R. Prothero and R. M. Schoch, eds., *The Evolution of the Perissodactyls.* Oxford: Oxford University Press.

Prothero, D. R., and R. M. Schoch, eds. *The Evolution of the Perissodactyls.* Oxford: Oxford University Press.

Schoch, R. M. 1989. "A review of the tapiroids." In D. R. Prothero and R. M. Schoch, eds., *The Evolution of the Perissodactyls.* Oxford: Oxford University Press.

Artiodactyls

Bonis, L. de, G. Koufos, and S. Sen. 1997a. "A giraffid from the middle Miocene of the island of Chios, Greece." *Palaeontology* 40:121–33.

Bonis, L. de, G. Koufos, and S. Sen. 1997b. "The Sanitheres (Mammalia, Suoidea) from the middle Miocene of Chios Island, Aegean Sea, Greece." *Revue paléobiologie de Genève* 16:259–70.

Bonis, L. de, G. Koufos, and S. Sen. 1998. "Ruminant (Bovidae and Tragulidae) from the middle Miocene (MN 5) of the island of Chios, Aegean Sea (Greece)." *Neue Jahrbuch geologische und paläontologische Abhandlungen* 210: 399–420.

*Crégut, E., and C. Guerin. 1996. "Famille des Bovidae." In C. Guerin and M. Patou-Mathis, eds., *Les Grands Mammifères Plio-Pléistocènes d'Europe.* Paris: Masson.

Delpech, F., and C. Guerin. 1996. "Famille des Cervidae." In C. Guerin and M. Patou-Mathis, eds., *Les Grands Mammifères Plio-Pléistocènes d'Europe.* Paris: Masson.

Fortelius, M., J. van der Made, and R. L. Bernor. 1996. "Middle and late Miocene Suoidea of central Europe and the eastern Mediterranean: Evolution, biogeography, and paleoecology." In R. Bernor, V. Fahlbusch, and W. Mittman, eds., *The Evolution of Western Eurasian Neogene Mammal Faunas.* New York: Columbia University Press.

Franzen, J. L., and G. Richter. 1992. "Primitive even-toed ungulates: Loners in the undergrowth." In S. Schaal and W. Ziegler, eds., *Messel: An Insight into the History of Life and of the Earth.* Oxford: Clarendon Press.

Gentry, A. W., and J. J. Hooker. 1988. "The phylogeny of the Artiodactyla." In M. J. Benton, ed., *The Phylogeny and Classification of the Tetrapods.* Vol. 2, *Mammals.* Oxford: Clarendon Press.

Gentry, A. W., G. E. Rössner, and E. P. J. Heizmann. "Suborder Ruminantia." In G. E. Rössner and K. Heissig, eds., *The Miocene Land Mammals of Europe.* Munich: Pfeil.

Ginsburg, L., J. Morales, and D. Soria. 1994. "The ruminants (Artiodactyla, Mammalia) from the Lower Miocene of Cetina de Aragón (province Zaragoza, Aragón, Spain)." *Proceedings of the Koninklijke Nederlandse Akademie van Wetenschappen* 97:141–81.

Heizmann, E. P. J. 1999. "Family Cainotheriidae." In G. E. Rössner and K. Heissig, eds., *The Miocene Land Mammals of Europe.* Munich: Pfeil.

Hünermann, K. A. 1999. "Superfamily Suoidea." In G. E. Rössner and K. Heissig, eds., *The Miocene Land Mammals of Europe.* Munich: Pfeil.

*Köhler, M. 1993. "Skeleton and habitat of recent and fossil ruminants." *Münchner Geowissenchaften Abhandlungen*, A 25:1–88.

Leinders, J. 1984. "Hoplitomerycidae fam. nov. (Ruminantia, Mammalia) from Neogene fissure fillings in Gargano (Italy)." *Scripta Geologica* 70:1–51.

Made, J. van der. 1994. "Suoidea from the Lower Miocene of Cetina de Aragón (Spain)." *Revista española de paleontología* 9:1–23.

Made, J. van der. 1996. "Listriodontinae (Suidae, Mammalia), their evolution, systematics and distribution in time and space." *Contributions to Tertiary and Quaternary Geology* 33:3–160.

Made, J. van der. 1999. "Superfamily Hippopotamoidea." In G. E. Rössner and K. Heissig, eds., *The Miocene Land Mammals of Europe.* Munich: Pfeil.

Made, J. van der, and J. Morales. 1999. "Family Camelidae." In G. E. Rössner and K. Heissig, eds., *The Miocene Land Mammals of Europe.* Munich: Pfeil.

Morales, J., L. Ginsburg, and D. Soria. 1986. "Los Bovoidea (Artiodactyla, Mammalia) del Mioceno inferior de España: Filogenia y biogeografía." *Paleontologia i evolució* 20:259–66.

Moyà-Solà, S. 1986. "El género *Hispanomeryx* MORALES: Posición filogenética y sistemática." *Paleontologia i evolució* 20:267–88.

Moyà-Solà, S. 1987. "Los rumiantes (Cervoidea y Bovoidea, Artiodactyla, Mammalia) del Ageniense (Mioceno inferior) de Navarrete del Río (Teruel, España)." *Paleontologia i evolució* 21:247–70.

Thomas, H. 1984. "Les Giraffoidea et les Bovidae Miocènes de la formation Nyakach (Rift Nyanza, Kenya)." *Palaeontographica* 183:64–89.

Vislobokova, V. 1992. "Neogene deer in Eurasia." In J. Agustí, ed., "Global Events and Neogene Evolution of the Mediterranean." *Paleontologia i evolució* 24–25:155–72.

Proboscideans, Hyraxes, and Archaic Ungulates

Franzen, J. L., and H. Haubold. 1986. "Ein neuer Condylarthre und ein Tillodonter (Mammalia) aus dem Mitteleozän des Geiseltales." *Palaeovertebrata* 16:35–53.

Godinot, M., T. Smith, and R. Smith. 1996. "Mode de vie et affinités de *Paschatherium* (Condylarthra, Hyopsodontidae) d'après ses os du tarse." *Palaeovertebrata* 25:22–242.

Göhlich, U. B. 1999. "Order Proboscidea." In G. E. Rössner and K. Heissig, eds., *The Miocene Land Mammals of Europe.* Munich: Pfeil.

Heissig, K. 1999a. "Order Tubulidentata." In G. E. Rössner and K. Heissig, eds., *The Miocene Land Mammals of Europe.* Munich: Pfeil.

Heissig, K. 1999b. "Suborder Hyracoidea." In G. E. Rössner and K. Heissig, eds., *The Miocene Land Mammals of Europe.* Munich: Pfeil.

Koenigswald, W. von. 1992. "The arboreal *Kopiodon,* a relative of primitive hoofed animals." In S. Schaal and W. Ziegler, eds., *Messel: An Insight into the History of Life and of the Earth.* Oxford: Clarendon Press.

Koenigswald, W. von. 1999. "Order Pholidota." In G. E. Rössner and K. Heissig, eds., *The Miocene Land Mammals of Europe.* Munich: Pfeil.

Kotsakis, T. 1987. "Neogene biogeography of Hyracoidea (Mammalia)." *Annales Instituti Geologici Publici Hungarici* 70:477–81.

Lister, A., and P. Bahn. 1994. *Mammoths.* New York: Macmillan.

Pickford, M., S. Moyà-Solà, and P. Mein. 1997. "A revised phylogeny of Hyracoidea (Mammalia) based on new specimens of Pliohyracidae from Africa and Europe." *Neue Jahrbuch geologische und paläontologische Abhandlungen* 205:265–88.

Shoshani, J., and P. Tassy, eds. 1996. *The Proboscidea: Evolution and Palaeoecology of Elephants and Their Relatives.* Oxford: Oxford University Press.

Storch, G., and G. Richter. 1992a. "The ant-eater *Eurotamandua*: A South American in Europe." In S. Schaal and W. Ziegler, eds., *Messel: An Insight into the History of Life and of the Earth.* Oxford: Clarendon Press.

Storch, G., and G. Richter. 1992b. "Pangolins: Almost unchanged for 50 million years." In S. Schaal and W. Ziegler, eds., *Messel: An Insight into the History of Life and of the Earth.* Oxford: Clarendon Press.

Tassy, P., and J. Shoshani. 1988. "The Tethytheria: Elephants and their relatives." In M. J. Benton, ed., *The Phylogeny and Classification of the Tetrapods.* Vol. 2, *Mammals.* Oxford: Clarendon Press.

INDEX

Numbers in italics refer to pages on which illustrations appear.

Aardvarks, 187, 212

Aboletylestes, 17

Aceratherines: Miocene, early, 90, 92, 111, 113; Miocene, late, 162, 185; Miocene, middle, 134; Pliocene, 212

Aceratherium: A. incisivum, 157, *158*, 185; Miocene, 111, 157

Acerorhinus, 162

Acheulian Industry, 276

Acinonyx pardinensis, 213, *222–23*, 223, 226, 231, 252

Acomys, 44

Acteocemas, 116

Adapids, 34–35, 53

Adapis: A. parisiensis, *55*, 55–56; Eocene, 53, 55

Adapisorex, 13

Adapisoricids, 7, 17, 25

Adapisoriculus, 17

Adcrocuta: A. eximia, *161*; Miocene, 174, 193, 208

Adelomyarion, 87, 98

Adunator, 17

Aegyptopithecus, 138, 139

Aenigmavis, 48

Africa: camels in, 208–9; creodonts in, 122; didelphids in, 25; hominids in, 167; humans, dispersal from, 244, 246; listriodontines in, 131; Messinian, 208; "modern" humans in, 276–77; Pleistocene, 253, 280; Pliocene glaciation period, 230, 231; primates in, 138; rhinos in, 91, 185; tethytheres in, 27

African–Iberian corridor, Eocene, 50

Agerinia, plate 1

Agnotherium: A. antiquus, 159; *A. grivensis*, 105, 137, 159

Agriotherium, 207, 214

Ailuravids, 45, 51, 53, 69

Ailuravus, 31, 45, 53

Albanensia, *121*, 142, 170

Albanohyus, 140

Alboran Microplate, 204

Alcelaphines, 197

Alces: A. alces, 250, 265; *A. latifrons*, 247, 250

Alephis, 212

Algeripithecus minutus, 138

Alicornops: A. alfambrensis, 185; *A. simorrense*, 142

Alilepus, 154, 156

Alligators, 20, 43, 48

Allognathosuchus, 20, 43, 48

Allohyaena, 138

Allometric evolution, 267
Allophaiomys: A. deucalion, 241; *A. plio-caenicus,* 241–42, 251; *A. vander-meuleni,* 241
Almogaver, 50
Alsaticopithecus, 44
Altomiramys, 98
Alveolines, 2
Amaranthaceae, 230
Amebelodon, 106
Ampelomeryx, 117, plate 6; *A. gins-burgi, 117*
Amphictis, 88, 90
Amphicynodon: A. leptorhynchus, 81, 82–83; Oligocene, 80
Amphicyon, 136; *A. castellanus,* 159; *A. giganteus,* 99, 102, 104, *104,* 137; *A. gutmanni,* 173; *A. longira-mus,* 88; *A. major, 136,* 137, 159, 173; *A. pannonicus,* 173; Miocene, 159, 173, plate 6
Amphicyonids: Eocene, 52; Miocene, early, 99, 102, 104–5, 119; Mio-cene, late, 154, 159, 173; Miocene, middle, 135, 137, 145; Oligocene, 87–88, 92, 99; teeth of, 87–88, 99, 102, 104–5, 137
Amphilemur, 44
Amphilemurids, 44
Amphimerycids, 63
Amphimeryx, 63
Amphimoschus, 203
Amphiperatherium, 25, *26–27,* 69, 135
Amphiprox, 159, 170
Amphitragulus, 97
Amynodonts, 70
Anancus: A. arvernensis, 205; Mio-cene, 205; Pliocene, 211–12, 234, plate 12
Anapithecus, 139, 164
Anchitherines, 110, 157
Anchitherium: A. aurelianense, 112; Mio-cene, 110, 119, 152, 157, plate 5
Ancylotherium, 190; *A. pentelecicum, 144*
Andegameryx, 116
Ankarapithecus: A. meteai, 164; Miocene, 147, 164, 166–67, 169, 174–75

Anoplotherids, 61, 72
Anoplotherium, 61
Antarctic Ice Sheet, 151, 204, 229
Antarctica, separation of Australia from, 67
Anteaters, 47–49, *49*
Antelopes: dwarf, 197; Holocene in-terglacial period, 279; Miocene, 178; Pleistocene, 251, 265; Plio-cene glaciation period, 221; saiga antelope, 265; spiral-horned, 180, 205, 212
Anthracobune, 59
Anthracoglis, 195, 197
Anthracotheres: and African–Eura-sian exchange, 106; Eocene, 59, 61; Miocene, 105, 106, 197; Oligo-cene, 61, 73, 85, 105
Anthracotherium, 73, 74
Anthropoids, 138–40
Antilopines, 130
Antlers: *Alces latifrons,* 250; *Croizeto-ceros,* 205, 212; *Dicroceru*s, 141; *Eu-cladoceros,* 234; *Euprox,* 141, 185; *Libralces gallicus,* 238, *239,* 250; *Lu-centia,* 185; of megacerines, 247, 250; *Megaloceros giganteus,* 267–68, *270; Pliocervus,* 212; *Procapreolus,* 185; structures of, 116
Ants, giant, 43
Apatemyids: Eocene, 25, 45, 53; Oli-gocene, 69; origins of, 6; Paleo-cene, 17
Apatemys, 25
Apes, 139, 148–49, 251
Apidium, 138
Aplodontia rufa, 78
Aplodontids, 78–79
Apocricetus, 218, 220
Apodemus, 218
Archaeomys, 78
Archaeonycteris, 32, 45
Archaeotherium, 74, 77
Arcius, 25
Arctocyon: A. primaevus, 11, 11–12, *12;* Paleocene, 10, 12
Arctocyonides, 12
Arctocyonids, 7, 10, 17, 25, 50
Arctoids, 80

Arctoparamys, 21
Arfia, 30
Armantomys, 135
Artemisia, 230, 253
Artiodactyls: archaic, 141; bunodont, 73; Eocene, 40, 47, 59–65; Miocene, 96, 113, 17; origins of, 50; and Pleistocene megafaunal extinction, 280
Arvernoceros, 226; *A. ardei*, 221, 231
Arvicolids: on Mediterranean islands, 262; Miocene, 173; Oligocene, 80; Pleistocene, 238, 241, 244, 262; Pliocene, 218, 219, 220
Asiatosuchus, 19–20, 43, 48
Astralagus, 36, 40
Atalonodon, 51
Atavocricetodon, 80
Atlantoxerus, 135
Atractosteus, 43
Aumelasia, 47
Aureliachoerus, 96, 132
Aurochs, 242, 254
Australopithecids, 148
Australopithecus afarensis, 148
Austroportax, 129–30

Baboons, 251
Bachitherium, 73; *B. insigne*, 73, 74
Badenian salinity crisis, 127
Badgers, 123, 162
Baluchithere, 70
Barbourofelis, 156
Barranco León (Spain), 252
Baryphracta, 43, 48
Barytheres, 108
Bats: Eocene, 31–33, 45; Miocene, 99, *100–103*
Bavarictis, 88, 90
"Bear-dogs," 52, 56, 87
Bears: brown, 207, 216, 250, 280; cave, 270, *272, 273,* 273–74, 279; Eurasian, 250; Miocene, 119, 159; Pleistocene, 250; Pliocene, 214, 216
Beavers, 78, 145, 170, 218
Bedenomeryx: B. truyolsi, 97; Miocene, 91, 118

Beetles, at Messel lake, 43
Bergertherium, 134
Bergisuchus, 49–50
Bernor, Ray, 128
Berruvius, 15
Betic Corridor, 204, 208
Birds, in Messel forest, 48
Birgerbohlinia, 182, plate 10; *B. schaubi, 182–83*
Bison, 242, 250; *B. menneri*, 250, *255; B. priscus*, 254, *255*, 265, *271; B. schoetensacki*, 250, 252, 253; *B. voigtstedtensis*, 256, plate 15; Holocene interglacial period, 279
Bjornkurtenia, 218
Blackia, 144
Blainvillimys, 53, 78, 85
Boars, 140, 234, 247, *271*
Body size, 195
Bohlinia, 182; *B. attica, 184, 186*
Boids, 43
Bone crushing, 174
Bos, 221; *B. primigenius*, 242, 254
Boselaphines: Miocene, 129, 134, 140, 170, *172*, 180, *181*, 205; Pliocene, 212
Boselaphus, 170; *B. tragocamelus*, 129
Bothriodon, 73
Boule, Marcellin, 276
Bovids: during glaciation, 221; Miocene, 118, 129–30, 140, *172*, 179, 180, *181*, 197, 208; ovibovine, 180, 182, 242, 251, 253, 265, 238; Pleistocene, 254, 262; Pliocene, 212, 221, 226, 238; rupicaprine, 238, *240;* teeth of, 263; tragelaphine, 251
Bowfins, 42–43
Brachydiceratherium, 91, 113
Brachyodus, 106
Brachypotherium, 134, 141–42, 179, 212
Braincase: Ceprano material, 255; of *Homo ergaster*, 244–45; of "modern" humans, 276; of Neanderthals, 274, 276; of *Ouranopithecus macedoniensis*, 167, *168*; Steinheim skull, 258; Swanscombe skull, 258
Bramatherium, 154

Bransatoglis, 85, 98, 170
Bubalus: B. bubalis, 255; *B. murrensis*, 254–55
Budorcas, 182; *B. taxicolor*, 180
Bunodontia, 47
Bunolistriodon, 113, 134, 140, plate 5, plate 7
Bunoselenodont forms, 72
Buxolestes, 6, 45
Byzantinia, 135, 187

Cadurcotherium, 70
Caenomeryx, 73
Cainotherids: Eocene, 61, 63; Miocene, 96, *97;* Oligocene, 73
Cainotherium, 63, 96, 141, plate 3, plate 4; *C. laticurvatum, 97*
Calvarium. *See* Braincase
Camelidae, 208
Camels, 208–9
Can Llobateres (Spain), 148, 157, 159, *165, 166*
Canids: Messinian, 207–8; Miocene, 193, 207; Pleistocene, 250; Pliocene, 213, *234;* Turolian, 207
Canine shear-bite, *84–85*
Canis: C. cipio, 193, 207; *C. etruscus, 234–35*, 236, 238, 242, 244, 250, 251; *C. falconeri*, 236–37; *C. lupus, 104*, 236, 250; *C. mosbachensis*, 250, 252, 256; Miocene, plate 11; Pliocene, 236
Cannon bone, 63, 65
Cantabrotherium, 59
Canthumeryx, 116
Cantius, 35
Capra ibex, 265
Caprines, 182, 253
"Carnassials," 29
Carnivores: Eocene, 28, 56; during glaciation, 222–23, 225, 226; Messinian, 207; Miocene, 104, 119–23, 135, 137, 145, 159, 179, 187, 199, 207; Oligocene, 80–83; Paleocene, 21–22; Pleistocene, 242, 250; Pliocene, 222–23, 225, 226, 237–38; teeth of, 29, 30–31
Carpathian Foredeep, 123
Carpolestids, 17

Castillomys, 218
Castor fiber, 218, 256
Castorids, 78, 142
Catarrhines, archaic, 138, 139
Cebochoerids, 47, 59
Celadensia, 218
Cephalogale, 88
Ceprano (Italy), 255
Ceratomorphs, 38
Ceratotherium, 212; *C. neumayri*, 185, *191; C. simum, 191*; Miocene, 162, 185
Cercopithecids, 251
Cercopithecine monkeys, 216, *217*, 218
Cerro Batallones (Spain), 155, *155*
Cervalces, 238
Cervids: dwarf, 261; during glaciation, 221, *224*, 226; Miocene, 116, 141, *141*, 159, 170, 185, 205; Pleistocene, 242, *243*, 244, 247, 250, 261; Pliocene, 212, 221, *224*, 226, 238, *239*
Cervoids, 92, 134, 170
Cervus, 221; *C. cusanus*, 221; *C. elaphus, 214*, 256; *C. elaphus acoronatus*, 247; *C. perrieri*, 221
Cetaceans, 13
Chalicomys, 144–45, 170
Chalicotheres, 110, 180
Chalicotherids: Eocene, 38; Miocene, *143, 144*, 156; Oligocene, 72
Chalicotherium: C. grande, 143, 144; Miocene, 142, 156, 180, plate 8
Chambered stomach, 64
Chamois, 265, *271*
Chasmaportetes, 213; *C. lunensis*, 213, *214–15, 216*, 222, 237
Chasmotherium, 51
Chatelperronian Industry, 278
Cheetahs, 223
Cheirogaster perpiniana, 218, 234, plate 12
Chelonians, 234
Chevrotains, 64
Chilotherium, 162, 185
Chionomys, 251
Chiromyoides, 17
Chiropterans, 31–32

Choenopodiaceae, 230

Choeropotamids, 47, 59

Choeropsis liberiensis, 261

Cincamyarion, 87

Circamustela, 162

Cistus, 230

Climacoceras, 116

Climate: Miocene, 93, 95, 124–28; Olduvai geomagnetic epoch, 241; Pleistocene, 241, 242, 246, 252, 263–64; Pliocene, 211, 218, 220, 229–31

Cockroaches, at Messel lake, 43

Coelodonta antiquitatis, *249*, 265, *268, 271*

Colobine monkeys, 175, 216, 234

Colugo, 34

Condylarths: Eocene, 25, 27, 28, 46, 50; extinction of, 50; limb articulation of, 7; in Messel forest, 46; Paleocene, 9, 10–13

Conohyus, 132, 134, 140, 170

Coryphodon, 21, 23, 28, *29*

"Cotton rats," 173

Cremohipparion, 185; *C. mediterraneum, 186, 188–89, 190; C. periafricanum,* 185, *188*

Creodonts: Eocene, 28–29, 56; Miocene, 104, 122–23; Oligocene, 69, *70*, 122

Cricetids: Miocene, 98, 114, 135, 142, 170, 187, 195, 199; Oligocene, 80; in Oligocene–Miocene transition, 98; Pliocene, 218, 220

Cricetines, 187

Cricetodon, 114, 115, 135

Cricetulodon, 170, 173

Cricetulus migratorius, 80

Cricetum vacuum, 114

Cricetus cricetus, 80

Criotherium, 182

Crocodiles: archaic, 8–9; Eocene, 43, 48–49; eusuchian, 18–20; in Messel forest, 48–49; Pliocene, 218, 234

Crocuta, 174; *C. crocuta*, 213, 251, 253, 256; *C. crocuta spelaea*, 268, *271*

Croizetoceros, 205, 212, 221; *C. ramosus*, 221, *224*, plate 13

Cueva Victoria (Spain), 251

Cuisitherium, 42

Cuon: C. alpinus, 250–51; *C. stehlini*, 250, 252

Cuvier, Georges, 206

Cyclurus, 42

Cymbalophus, 36

Cynelos, 88, 99; *C. helbingi*, 99, 137; *C. schlosseri*, 99, 137

Cynocephalus, 34

Cynodictis, 56, 80

Cynohyaenodon, 56

Dacritheryum, 61

Dacrytherids, 61, 72

Dama, 251; *D. carburangelensis*, 261; *D. dama*, 244; *D. nesti*, 244; *D. vallone-tensis*, 244, 252, 256, plate 16

Daphoenus vetus, 88

Darwin, Charles, 274

Daubentonia, 45

Debruijnimys, 218

Decennatherium, 154, 156, 182

Deer: dwarf, 261; giant, 267, *268, 269, 270;* megacerine, 221, 242, 247; Miocene, 116; mouse, 113; Pleistocene, 242, *243,* 244, 260, 267, *268, 269, 270;* Pliocene, 221, 233–34; red, 247, *271*

Deinogaleryx: D. koenigsvaldi, 200–201; Miocene, 199, 202

Deinotheres, 108, 156, 180

Deinotherium: D. giganteum, 111; D. gigantissimum, 180; Miocene, 108, 110, 156, 179, plate 8, plate 10

Democricetodon, 114, 115

Deperetomys, 135

Dermopterans, 31–34

Desert–savanna boundary, 230

Dholes, 250

Diaceratherium, 113, 141; *D. aurelianense, 96; D. brachypus*, 113; Miocene, plate 6

Diacodexis, 40, 42, 47, 50

Diaphyodectes, 17

Diatryma: D. geiselensis, 48; Paleocene, *12,* 18

Diceratheres, 70

Dicerorhinus, plate 5; *D. sumatrensis*, 141–42

"Dicerorhinus" steinheimensis, 142, 169

Diceros, 147, 185, 212; *D. bicornis*, 147; *D. pachygnatus*, 162; *D. primaevus*, 185

Dichobune, 59, 72

Dichobunids, 42, 47, 59, 72

Dicrocerus, 116, 141

Didelphoids, 9, *10*, 24–25, 50, 69

Dinocrocuta: D. gigantea, 163; Miocene, 137–38, 162, 193

Dinofelis, 207, 213, 223, plate 12

Diplobune, 61, 72

Diplocynodon, 20, 43, 48

Dissacus: D. europaeus, 14, 14–15; Paleocene, 12–13, 17

Djebel Irhoud-1 (Morocco), 256, 258

Dmanisi (Georgia), 244, 252

Dolichopithecus, plate 12; *D. rusciniensis*, 218

Doliochoerus, 77

Donrussellia, 35

Dorcatherium, 113–14, 159, 180, plate 7

Dormice, 51, 85, 195, 199, 262; on Mediterranean islands, 262; Miocene, 115, 135, 179, 187, 202; Pleistocene, 260–62

Dremotherium, 97, 98; *D. cetinensis*, 98

Dryomys, 85, 187

Dryopithecids, 197

Dryopithecus: D. crusafonti, 164; *D. fontani*, 147; *D. hungaricus*, 164; *D. laietanus, 165, 166;* Miocene, 145, 147–49, 164, 169, 174–75, 178, 193

Dwarfism, 197, 260–62

Dyrosaurids, 8–9

Early Miocene, 95–123: carnivores of, 119–23; climate of, 93, 95; mammals of, 95–105; proboscidean event in, 105–10

Early Pleistocene, 241–52

Early Pliocene, 211–19

Early Vallesian climax, 156–62

East Antarctic Ice Sheet, 229

Eastern Antarctic Ice Sheets (EAIS), 124–27

Ebro Basin (Spain), 68

Eburonian stage, 241, 246

Echinosorex, 199

Ectolophes, 135

Edentates, 48, 280

Eemian interglacial phase, 264

Egyssodon, 69, plate 4

Eivissia, 261

Elands, 251

Elasmotherines, 134, 162

Elephantids, 231, *232*

Elephants: dwarf, 260–61; Pleistocene, 247, 260, 264; straight-tusked, 247, 264

Elephas: E. antiquus, 247, 253, 260, 261, 264; *E. cretica*, 261; *E. cypriotes*, 261; *E. falconeri*, 260; *E. melitensis*, 261; *E. mnaidrensis*, 260–61

Elfomys, 53

Eliomys, 187; *E. quercinus*, 262

Elk, 238, 242, 247, 250, 265, 267

Elomeryx, 73, 74, plate 3

Encephalic volume, of Neanderthal, 274, 276

End Cretaceous Mass Extinction, 2, 3

Entelodon, 7, 74, 77, 85; *E. deguilhemi, 75, 76–77, 86–87; E. magnum,* 77

Entelodontids, 7, 74, 77

Eobison, 242, 250

Eocene epoch, 23–65; forests of, 23; landmasses of, 24, 28, 42, 50, 52–53, 59; late, 50–65; mammals of, 30–35; Messel forest in, 42–50; middle, 42–50; North America in, 28–30, 31, 37, 38; temperature in, 23; ungulates of, 36–42

Eogliravus, 51

Eomanis: Eocene, 40, 41, 47–48; *E. waldi, 40–41*

Eomellivora, 162

Eomuscardinus, 170

Eomyids: Miocene, 98, 114, 135, 142, 144; Oligocene, 79–80, 98; teeth of, 79, 80, 98–99

Eomyops, 144

Eomys, 80, 85

Eotalpa, 78

Eotapirus, 95

Eotragus, 118–19, 129, 140, plate 6

Ephedra, 230

Epiaceratherium, 72, 90

Equids: Eocene, 36; Miocene, 152; Pliocene, 232–33; teeth of, 37

Equus: E. altidens, 233, 247, *248*, 252, 253, 256, plate 14; *E. bressanus*, 251; *E. caballus*, 247, 253, *271; E. germanicus*, 253; *E. granatensis, 248; E. hydruntinus*, 233, 253; *E. mosbachensis*, 253; *E. stehlini*, 247; *E. stenonis*, 233, *233*, 242, 244, 247, 251, 252, plate 13; *E. sussenbornensis*, 247; Pleistocene, 253; Pliocene, 232, 234

Equus–Mammuth event, 231–36

Erinaceids, 7, 199, *200–201*

Erinaceomorphs, 52

Esthonyx, 21, 28, 50

Eucladoceros, 233–34, 242; *E. giulii*, 242, *243*, 244, 256, plate 15; *E. senezensis, 214*, 233–34, 242

Eucricetodon, 87, 114

Eucyon, 207

Eumaiochoerus, 197, 198

Eumyarion, 114, 115, 170

Euprimates, 34

Euprox, 116, 141

Eurasia, Miocene, 110–15

Euroamphicyon olisiponensis, 104, 137

European Archipelago, 68

Europolemur, in Messel forest, 45–46

Eurotamandua: Eocene, 48, 49; *E. joresi, 49*

Euroxenomys, 145, 170

Eusmilus: E. bidentatus, 84–85; Oligocene, 81, *86–87*

Evolution, allometric, 267

Face, of hominoids, 164, 166, 167, 274

Felids: Miocene, 119–21, 145, 154–56, 188; Pliocene, 223

Felines, 190

Felis attica, 190, *196*, 207

Fish, in Messel pit, 42–43

Flora: Eocene, 23–24; in Greek-Iranian Province, 128–38; Holocene interglacial period, 279; in Messel forest, 43; Miocene, 94–95; of Oak-Laurel Forest, 95; Pleistocene, 247; Pliocene, 211, 230–31; during Vallesian Crisis, 176–78

Flying squirrels, 31; Miocene, 135, 142, 170; Pliocene, 218, 220

Forelimbs: entelodontids, 77; *Machairodus*, 155

Forests, Eocene, 23

Forstercooperia, 69

Foxes, 207, 213, 236, 250

Franzenium, 59

Fuentenueva (Spain), 252

Gallogoral meneghini, 238, *240*

Gargano fauna, 199–203

Gars, 42–43

Gastornis, 18

Gazella, 130, 212, 238

Gazelles: Miocene, 130, 180, 197, 205; Pliocene, 212, 238

Gazellospira: G. torticornis, 221, *222–23*, plate 13; Pleistocene, 221, 238

Gelocidae, 64, 91

Gelocus, 73, 91

Genets, 123

Gentrytragus, 130

Georgiomeryx, 131

Gerbils, 218, 244

Gervais, Paul, 198

Gibbons, 139

Gibraltar, 210

Giraffa, 115, 182; *G. camelopardalis, 184*

Giraffes, *184*

Giraffids, 115–16; Miocene, late, 154, 182, *182–83, 186*, 197; Miocene, middle, 145; Pliocene, 212, 221

Giraffoids, 115, 118

Giraffokeryx, 130, 131, 145

Glaciation, 279; Eburonian stage, 241, 246; Holocene interglacial period, 279–80; Pleistocene, 241, 252, 260, 263–74; Pliocene, 219–20, 229–31

Gliravus, 51, 78, 85

Glirids: Eocene, 51; Miocene, early, 98, 115, 170; Miocene, late, 170, 187, 195, 202; Miocene, middle, 135, 142, 144, 170; Oligocene, 78, 85; teeth of, 51, 85, 135

Glirudinus, 98

Glirulus, 144, 187

Glis, 187

Glivarus, 98
Gnus. *See* Wildebeests
Goats, 182, 197, 242
Gomphothere Bridge, 105, 114, 122, 138
Gomphotheres: Miocene, 106, 156, 205, 206, 231; Pliocene, 211; teeth of, 106, 108
Gomphotherium: G. angustidens, 107, 111; G. steinheimensis, 107; Miocene, 106, 145, 206, plate 5, plate 6
Gondawanaland, 67
Gould, Stephen Jay, 267
Graecopithecus. See *Ouranoupithecus*
Graecoryx, 170
Gran Dolina, 256, *256*
"La Grand Coupure," 68–69, 80
Greek-Iranian Province: Miocene, early, 128–38, 139, 140; Miocene, late, 154–55, 162, 164, 167, 169, 175, 185
Greenland Ice Sheet, 229
Griphopithecus, 139, 149, 169
Gruiforms, 48
Guadix-Baza Basin (Spain), 252
Gulo gulo, 271

Hadrictis, 162
Haenodon, 56
Hainina, 4, 24
Hallensia, 36, 46
Hamsters: Miocene, 115, 170, 187, 199; Oligocene, 80; Pliocene, 218
Hand anatomy: *Dryopithecus,* 148; *Heterohyus,* 45
Haplobunodontids, 47, 59
Haplocyonoides, 99
Haplogale, 87
Hartebeests, 197
Hassianycteris, 45
Hattomys, 199, 201, 202
Hedgehogs, 78, 199, *200*
Helladotherium, 182
Hemicyon: H. goeriachensis, 159; *H. sansaniensis, 120;* Miocene, 119, 145, 159, 173, plate 5
Hemiones, 247

Hemitragus: H. albus, 242, 251, 252; *H. bonali,* 253; *H. jemlahicus,* 242; Pleistocene, 253
Herbivores, Miocene, 114, 129, 152
Herpestids, *124*
Heterocricetodon, 87
Heterohyus, 45, 53
Heteroprox, 141, *141; H. larteti, 148*
Hexaprotodon, 208, 212
Hidaspitherium, 154
Hipparion: H. antelopinum, 185; *H. capmbelli,* 185; *H. concudense,* 185; *H. crassum,* 212–13; *H. fietrichi,* 185; *H. fissurae,* 213; *H. gettyi,* 185; *H. gromovae, 188; H. melendezi,* 185; *H. primigenium,* 152, *152–53; H. prostylum,* 185; *H. rocinantis,* 232–33; Miocene, 151, 152, 154, 156, *160,* 178, plate 9, plate 11; Pliocene, 232, 234, plate 12
Hipparionine horses: Miocene, 151–56, 185, *186, 188;* Pliocene, 212–13
Hippomorphs, 36
Hippopotamus: dwarf, 261; of Mediterranean, 261; Miocene, 208; Pleistocene, 244, *244–45,* 260, 264
Hippopotamus: H. amphibius, 244, 261, plate 16; *H. creutzburgi,* 261; *H. major,* 244, *244–45,* 251, 252, 261; *H. melitensis,* 261; *H. pentlandi,* 261
Hipposiderids, 33, 99, *102, 103*
Hipposideros bouziguensis, 99, *102, 103*
Hippotherium: H. primigenium, 188; Miocene, late, *160,* 185, plate 10
Hippotragini, 208
Hispanochampsa, plate 3, plate 4
Hispanodorcas, 180, 205, 212, plate 11
Hispanomeryx, 131, 141, 159, 170
Hispanomys, 135, 187
Hispanotherium, 134, 178, plate 7
Holocene interglacial period, 279–80
Hominoids: Miocene, late, 162–69, 174–75, 178, 198; Miocene, middle, 139, 145, 147, 164, *165;* teeth of, 139, 167, 187
Homo: H. antecessor, 256, *256–57,* 258, 259; *H. erectus,* 246, 253, 256; *H. ergaster,* 244, 246, 256, 280; *H. heidel-*

bergensis, 258–60, *259*, 275; *H. neanderthalensis*, 258, 274, *275*, 276–79; *H. sapiens*, 280
Homoiodorcas, 130
Homotherium, 223, 225, 226; *H. crenatidens*, 226, 231, 242, 244, 251, 252; *H. latidens*, 227, 228
Hoplictis, 123
Hoplitomerycidae, 202
Hoplitomeryx, 202, 203; *H. matthei*, 203
Hoploaceratherium, 111, 142, plate 8; *H. tetradactylum*, 142
Horns: *Bison priscus*, 254; of bovids, 118; *Eotragus*, 118; *Gentrytragus*, 130; *Hoplitomeryx*, 202; *Leptobos*, 221; *Megalovis*, 221; of ovibovines, 180; *Parabos*, 205; *Prostrepsiceros*, 180; of rhinocerotines, 91; *Stephanorhincus hemitoechus*, 247; *Tethytragus*, 130; of tragelaphines, 251; *Tragoportax*, 170
Horses: Eocene, 36; hipparionine, 151–56, 185, *186*, *188*, 212–13; Holocene interglacial period, 279; Miocene, 110; in North America, 110; Pleistocene, 253, 253–54, *254*, *271*, 280; Pliocene, 232, 232–33, 233, *233*; stenonine, *233*, 247, 253; teeth of, 36–37
Huerzelerimys, 170, 187, 195
Humans: dispersal of, from Africa, 244, 246; domestication by, 281; extinction of animals and, 279–81; *Homo sapiens*, 280; Middle East succession of, 277; "modern," 276–79; Neanderthals, 258, 274, *275*, 276–79; Pleistocene, 255–60, 274–80; Pleistocene, early, 244, 246, 251–52; skulls of, from Africa, 276–77; in western Europe, 251–52. *See also* Homo
Hürzeler, Johannes, 198
Huxley, Thomas, 274
Hyaemoschus, 64, 114; *H. aquaticus*, 113
Hyaena hyaena, 213
Hyaenictis, 174
Hyaenictitherium, 193, 213
Hyaenidae, 173

Hyaenids: Miocene, 122–23, 137, 145, 159, *160*, *161*, 174, 191, 208; Pleistocene, 268; Pliocene, *214–16*; teeth of, 23, 145, 159
Hyaenodon: Eocene, 56, 65, plate 2; *H. dubius*, *70–71*; Oligocene, 69
Hyaenodontids, 65, *70*, 104
Hyaenodonts: Eocene, 30, 46, 52, 56; in Messel forest, 46; Miocene, 105; Oligocene, 69, 92, 105
Hyaenotherium: H. wongii, *161*; Miocene, 159, 193
Hyainailouros, 122–23; *H. sulzeri*, *104*
Hydropotes, 65
Hyenas: Pleistocene, 251, *271*; Pliocene, 213, *214–15*, 237, *237*; spotted, 213, 251, 268; striped, 213
Hylobatids, 138
Hylochoerus, *133*, 140
Hyopsodontids, 13, 27, 50, *58*
Hyopsodus, 27
Hyotherines, 96, 132, 140
Hyotherium, 96, 132, 140
Hypercarnivores, 81, 104, 137, 154, 159
Hypnomys, 261, 262
Hypsodontus, 130; *H. miocaenicus*, 130; *H. serbicum*, 130
Hypsodonty, 21, 78, 129, 130, 220, 240, 241, 251, 263
Hyrachius: Eocene, 38, 40, 46, 47; *H. minimus*, *40–41*; Oligocene, 69
Hyracodonts, 69, 90
Hyracotherium: Eocene, 36; *H. leporinum*, 36, 37; *H. pernix*, 36; *H. vulpiceps*, 36; origins of, 50
Hyraxes, 50, 157, 187
Hystrix, plate 10

Iberian corridor, Eocene, 50
Ibex, 265
Icaronycteris, 32
Ice sheets, 263–74; Holocene interglacial, 279–80; Miocene, 125, 151; Pleistocene, 252; Pliocene, 219–20, 229–31
Ictiocyon socialis, 105, 137
Ictitherium: I. viverrinum, *161;* Miocene, 159, 193, 208

Indarctos, 173, 199; *I. arctoides*, 173; *I. atticus*, 193, 207; *I. vireti*, 159, 173; Miocene, 159, 173; Pliocene, 214
Indricotheres, 70
Indricotherids, 69, *70*
Insectivores: Eocene, 25, 43, 52; Miocene, 170; Oligocene, 78; origins of, 7
Insular evolution, 260–62
Irish elk, 242, 267, *270*
Ischymomys, 173, 218
Ischyrictis, 123
Ischyromyids, 21, 31, 45, 51, 53, 69
Issiodoromys, 78, 85, 98, 114

Jaguars, 225, 238
Jaramillo epoch, 252
Java, *Pithecanthropus erectus*, 246
Jepsenella, 6, 17
Jirds, 265

Kenyapithecus, 139, 149
Keramidomys, 144
Kislangia, 238
Kogaionids, 4, 9, 24
Kogaionon, 4
Köhler, Meike, 198
Koobi Fora Formation, 231
Kopiodon, 27, 46
Kowalskia, 187, 195
Kritimys, 262
Kubanochoerus, 130, 131; *K. gigas*, *133*
Kudus, 251

Lagomerycids, 116, 141
Lagomeryx, 135, 141
Lagomorphs, 77–78, 114, 170, 195
Lagopsis, 114, 197
Landenodon, 25
Landmasses: Eocene, 24, 28, 42, 50, 52–53, 59; Miocene, 93, 105, 126–27, 134, 152, 204; Oligocene, 67–68, 83; Paleocene, 2
Langhian Transgression, 123–24, 126
Lantanotherium, 199
Lartetotherium sansaniense, 142, 169
Late Miocene, 151–210; early Vallesian climax in, 156–62; landmasses

of, 204; Mediterranean Sea in, 209–10; Messinian Crisis in, 204–9; Turolian stage of, 178–93; Tusco-Sardinian island in, 193–203; Vallesian Crisis in, 169–78, 193; Vallesian hominoid radiation during, 162–69
Late Pleistocene: climate of, 263–64; steppe-tundra biome of, 263–74, 279
Latest Paleocene Thermal Maximum, 23, 24
Laurisilva, 95
Le Vallonet (France), 252
Leinders, Joseph, 202
Leithia melitensis, 262
Lemmings, 265
Lemuriforms, 34
Lemurs, flying, 33–34
Leopards, 174, 251, 253, *271*
Leporids, 154
Leptacodon, 17
Leptadapis: Eocene, 53, 56; *L. magnus*, 57
Leptictidium, 5; Eocene, 44; *L. nasutum*, *44*
Lepticids, 5, 17, 51–52, 69
Leptobos: *L. elatus*, 231, plate 13; *L. etruscus*, 242; Pleistocene, 242; Pliocene, 221, 226
Leptoplesictis aurelianensis, 123, *124*
Lepus, 154
Libralces gallicus, 238, *239*, 250
Libyochoerus, 131
Ligerimys, 114
Ligeromeryx, 116
Limb structure: *Diacodexis*, 40; of paleomerycids, 118; of tillodonts, 21; of tragulids, 114; of ungulates, 36, 37
Limnonyx, 162
Lions, 253, 270, *271*
Liotomus, 9
Lisbon Basin (Portugal), 93
Listriodon, 140, 157, 159, 170, plate 8
Listriodontines, 113, 131–32, 134, 140
Locomotion: *Dryopithecus*, 149; of hipparionine horses, 152–53

Lophiaspis, 38, 72
Lophiobunodon, 59
Lophiodon, 38, 51, plate 1
Lophiodontids, 38, 46–47, 53, 56
Lophiomeryx, 73
Lophodonts, 78
Lorancahyus, 96–97
Lorancameryx, 115
Louisina, 13, 27
Lucentia, 185, plate 10
Lutra, 162
Lycaon pictus, 236
Lycyaena, 193, 208, plate 11
Lynx: L. issiodorensis, 213, *214; L. spe-
    laea*, 251, 252, 253; Pleistocene,
    256; Pliocene, 213
Lynxes, 253, 280
Lyptotiphla, 7

Macaca, 216; *M. prisca*, 216; *M. syl-
    vana*, 216, 247, 253
Macaques, 216, 247
Machairodontines: Miocene, 154–55,
    174, 187, 190, 207; Pleistocene,
    253; Pliocene, 213, 223; teeth of,
    187–89
Machairodus: M. alberdiae, 155; *M.
    aphanistus*, 155, *155*, 174, *192; M.
    copei*, 155; *M. giganteus*, 187, *192*,
    194–95, 207; *M. kurteni*, 207; *M.
    laskaveri*, 207; *M. romeri*, 155; Mio-
    cene, 154, 199, plate 9, plate 10,
    plate 11; Pliocene, 213
Macrocranion, 7, 43
Maltamys, 261
Mammals: arboreal, 34; Eocene, 30–
    35, 43–47; "flying," 31–34; glacia-
    tion and, 220–21, 231; in Greek-
    Iranian Province, 128;
    "insectivorous," 5–6; in Messel
    forest, 43–47; Miocene, 95–105,
    135, 170, 187, 195, 202, 205; Oli-
    gocene, 77–80; Paleocene, 8–17;
    placental, 5, 17, 25, 51–52, 69;
    Pleistocene, 252, 253, 260, 260–61,
    265, *271*, 280; Pliocene, 220, 221,
    231; Turolian stage, 178–93
Mammoths, 206; Holocene intergla-
    cial period, 279; Pleistocene, 280;
    steppe, 253–54, *254;* teeth of, 207,
    264; woolly, 264, *266–67*
Mammut: M. borsoni, 206; Miocene,
    206–7; Pliocene, 212
Mammuthus, 231–32, 234; *M. meri-
    odionalis*, 231–32, *232*, 242, 247,
    251, 252, 253, 254, plate 13; *M.
    primigenius, 232*, 264, *266–67; M.
    trogontherii*, 253–54, *254*, 264
Mangroves, 95
Marcetia, 162
Maremmia, 197
Marmota, 256
Marmots, 265
Marsupials: Eocene, 24–25; Miocene,
    135; Oligocene, 69; Paleocene, 9;
    Pleistocene, 280
Martens, 123
Martes munki, 123
Massillabune, 47
Massillamys, 45
Mastodons: Miocene, 106, *107*, 108,
    145, 206–7, *206*, 211; Pleistocene,
    280; Pliocene, 211, 231; zygodont,
    206, *206*, 211–12
Mediterranean Sea: during
    Messinian Crisis, 204–9; Pleisto-
    cene, 260–63; Pliocene, 209–10,
    220
Megacricetodon, 114–15, 135, 170
Megaloceros: M. cazioti, 261; *M. creten-
    sis*, 261; *M. giganteus, 214*, 242,
    267, 268–70; *M. giganteus antece-
    dens, 268–69; M. giganteus hiberni-
    cus, 270; M. savini, 214*, 242, 251;
    M. verticornis, 247, 252, 261; Pleis-
    tocene, 234, 267
Megalovis, 221, 226, 238
Megantereon: M. cultridens, 225, 226,
    251; *M. megantereon*, 226, 231, 244;
    Pleistocene, 253; Pliocene, 223,
    225, 226, 227
Megistotherium, 104
Meldimys louisi, 31
Meles meles, 252
Melissiodon, 87, 98, 114, 201
Meniscotherids, 13
Menoceras, 90, 91, 96, 111
Menoceratheres, 70, 72

Mentoclaenodon, 12

Merialus, 25, 27

Merychippus, 152

Mesaceratherium, 90–91, 96, 111, plate 5

Mesembriacerus, 182

Mesocarnivores, 88, 159

Mesocricetus auratus, 80

Mesohippus, 110

Mesomephitis, 152

Mesonychids, 12, 17

Mesopithecus: M. monspesulanus, 216, 218; *M. pantelecicum, 175; M. penteleci,* 187; Miocene, 175, 187; Pliocene, 234

Mesosuchians, 8, 49–50

Messel forest (Germany), 42–50

Messelbunodon, 47

Messinian Crisis, 204–9

Metailurines, 187, 207, 213

Metailurus: M. major, 187, 207; *M. parvulus,* 187, 207; Miocene, 187, 199; Pliocene, 213

Metaschizotherium, 110–11

Methriotherium, plate 4

Miacidae, 28, 30, 56

Miacis, 46

Mice, 280

Microadapis, 45

Microchoerids, 53–56

Microdyromys, 85, 98, 197

Microchoerus, 53

Microhyus, 27

Micromeryx, 131, *132,* 134, 159, 180, 202–3

Microparamys: Eocene, 31, 45, 51; *M. chandoni,* 31; *M. russelli,* 31

Microstonyx: M. major, 171, 205; Miocene, 170; Pliocene, 212

Microtia, 201–2

Microtocricetus, 173, 187

Microtodon, 219

Microtus, 251

Middle East corridor, Pleistocene, 246

Middle East succession, 277

Middle Miocene: climate of, 124–28; Greek-Iranian Province in, 128–38; landmasses of, 134; primates of, 138–40

Middle Miocene event, 123–28

Middle Pleistocene, 252–63; humans of, 255–60; Mediterranean Sea in, 260–63

Middle Pliocene Warming, 229

Mimomys, 220, 238

Miocene epoch: carnivores of, 119–23; early, 95–123; Eurasia in, 110–15; Gomphothere Bridge in, 105, 114, 122, 138; Greek-Iranian Province in, 128–38, 139, 140; ice sheets in, 125, 151; landmasses of, 93, 105, 126–27, 134, 152, 204; Langhian Transgression, 123–24; late, 151–210; mammals of, 95–105; Messinian Crisis in, 204–9; middle, 123–49; Middle Miocene Event, 123–28; proboscidean event, 105–10; temperature in, 93; Tusco-Sardinian island in, 193–203

Miodyromys, 98

Miohippus, 110

Miomachairodus pseudailuroides, 155

Miomephitis, 123

Miopetaurista, 170

Miotragocerus: Miocene, 129, 130, 134, 140, 180; *M. panonniae, 181*

Mixodectids, 6, 17

Moeritherium, 106

Moles, 52

Molossids, 99, *100–102*

Mongooses, *124*

Monkeys, 175, 216, *217,* 218, 234

Monshyus, 13, 27

Monterey Formation (California), 127

Monterey Hypothesis, 127

Moropus, 110

Moschids, *132,* 134, 170

Moschoids: Miocene, 97, 113, 141, 159, 180; Oligocene, 65, 91; teeth of, 98

Moschus, 65

Mouflon, 250

Mouillacitherium, 59, 61

Mousterian Industry, 276

Mouse deer, 113

Moyà, Salvador, 198

Multituberculates, 3–4, 9, 23, 24, 50

Murids: on Mediterranean islands, 262; Miocene, 170, 187, 195, 201–2; Pleistocene, 262; Pliocene, 218, 220
Mus musculus, 280
Muscardinus, 187
Musk ox, 180, 238, 265, 279
Mustelictis, 90
Mustelids, 8, 110, 123, 125, 159, 162, 174
Myocricetodontines, 135
Myoglis, 170
Myomimus, 85, 187
Myotragus, 197, 262–63; *M. antiquus*, 263; *M. balearicus*, 262–63; *M. batei*, 263; *M. kopperi*, 263; *M. pepgonellae*, 263; teeth of, 262–63

Nandinia binotata, 83
Nannopithex, 53
Neanderthals, 258, 274, 275, 276–79
Necrolemur: Eocene, 53; *N. antiquus*, 54
Neocometes, 114, 115
Neofelis nebulosa, 174
Neoplagiaulax, 9, 24
Neoteny, 80
Neotragines, 197
Nesiotites, 262
Nilgai, 129
Nimravides catacopis, 155
Nimravids: Miocene, 99, 119, 145, 173, 174, 190; Oligocene, 81, 87
Nimravus, 81, plate 4
Nisidorcas, 180, 205
North America: Eocene, 28–30, 31, 37, 38; equoid group in, 37; hipparionine horses in, 152–56; horses in, 110; lophiodonts in, 38
North Atlantic Ocean, Pleistocene, 252
North European Ice Sheet, 252, 264
Northern Hemisphere, glaciation in, 229–31, 264
Notochoerus, 132
Nyanzachoerus, 132
Nyctereutes, 208, plate 12; *N. donnezani*, 207, 213, 222; *N. megamastoides*, 222, 236; *N. procyonoides*, 207
Nyctitherids, 52, 53

Oak-Laurel Forest, 95
Occitanomys, 170, 187, 218, 220
Oceans: during Messinian Crisis, 209–10; Miocene, 126–27, 151; Pleistocene, 252–53; Pliocene, 218
Ochotona, 78
Ochotonids, 78, 114, 195
Oioceros, 180, 205
Okapis, 115
Olduvai geomagnetic epoch, 241
Oligocene epoch, 69–92; carnivores of, 80–83; landmasses of, 67–68, 83; mammals of, 77–80; ungulates of, 69–77, 90–92
Oligopithecus, 138
Omomyidae, 34, 35, 46
Opossums, 9–10
Oreopithecus, 193, 197, 198–99
Oriomeryx, 97, 118
Orohippus, 37
Orthaspidotherium, 13, 17–18
Orycteropus, plate 12; *O. afer*, 187; *O. depereti*, 212; *O. gaudryi*, 187
Oryctolagus, 154
Ossicones, 115, 118, 131, 154, *184*
Osteoborus, 213
Otters, 123, 162, 199
Ouranopithecus, 147, 167, 169, 175; *O. macedoniensis*, *168*
Ouzoceros, 180, 205
Ovibos: Miocene, 182; *O. moschatus*, 180, 265
Ovibovine bovids: Miocene, 180, 182; Pleistocene, 242, 251, 253, 265; Pliocene, 238
Ovis: O. ammon, 250, 253; *O. antiqua*, 250, 251, 252; Pleistocene, 253
Oxacron, 61
Oxen, wild, 254
Oxyaena, 29–30
Oxyaenids, 65

Pachycrocuta: P. brevirostris, 237, *237*, 238, 242, 251, 252; *P. perrieri*, 222, 226, 231, 237, 244; *P. pyrenaica*, 213, 222
Pachycynodon, 80
Pachynolophus, 56
Pachytragus, 182, 197
Pagomomus, 17

Palaeochiropteryx: Eocene, 32–33, 45; *P. tupaiodon, 32, 33*

Palaeochoerus, 77, 96

Palaeomastodon, 106

Palaeomerycids, 135

Palaeomeryx, 116–17

Palaeophyton, 43

Palaeoprionodon, 83; *P. lalandi, 89*

Palaeoryctes, 5

Palaeoryctids, 5, 17, 29

Palaeoryx, 182

Palaeosciurus: P. feignouxi, 98; *P. goti,* 98

Palaeotragus: Miocene, 116, 145, 147; *P. coelophrys,* 182; *P. moldavicus,* 182; *P. roueni,* 182

Paleictops, 17

Paleocene epoch, 1–22; Cretaceous inheritance of, 3–8; landmasses of, 2; Latest Paleocene Thermal Maximum in, 23, 24; mammals in Europe in, 8–17; nonmammalian predators of, 17–20; Paleocene–Eocene boundary in, 23; temperature in, 3

Paleochoerids, 77, 96, 113, 132, 134

Paleoloxodon antiquus, 232

Paleomerycids, 116–18, *117,* 131

Paleomoropus, 38

Paleonictis, 29–30

Paleoplayticeros, 141

Paleoryx pallasi, 190

Paleosciurus, 78, *95; P. goti,* 78

Paleotheres: Eocene, 37, 56, 59, 65; Oligocene, 69, 83, 85

Paleotherids, 37, 46, 56, 59

Paleotherium: Eocene, 56, plate 2; Oligocene, 69; *P. magnum,* 59, 60, *62–63, 64; P. medium,* 69; *P. muehlbergi,* 60

Paleotragines, 154, 182

Palm civets, 83

Paludolutra, 199

Paludotona, 195, 197

Panama, Isthmus of, 208–9

Pangolins, 47–48

Panthera: P. gombaszoegensis, 238, 251, 252, 256; *P. leo,* 253, 270; *P. leo spelaea, 271; P. onca,* 238; *P. pardus,* 251, 253, *271; P. spelaea,* 270; Pleistocene, 244

Pantodonts, 9, 20, 23, 28

Pantolestids, 5–6, 17, 27, 45

Parabos, 205, 212, 221

Paracamelus, 208, 212

Paraceratherium, 69–70, *70–71*

Parachleuastochoerus, 132, 140, 157, 159, 170

Paracricetodon, 87

Paracynohyaenodon, 56

Paradelomys, 53

Paradolichopithecus, 216; *P. arvernensis, 217*

Paraethomys, 218, 220

Paragale, 90

Paraglirulus, 170

Parahyaena, 174

Paralophiodon, 51

Paralutra, 123

Paramachairodus: P. ogygia, 174, 187; *P. orientalis,* 207; Pliocene, 213

Paramiacis, 56

Paranchilophus, 59

Parapithecus, 138

Parapodemus, 170, 187, 195

Paratapirus, 95

Paratethys Sea, 68, 123, 127, 204

Paratricuspiodon, 13

Parodectes, 46; *P. feisti, 26–27*

Paromomyids, 25

Paroxacron, 61

Paroxyaena, 56

Paroxyclaenids, 25, 46

Paroxyclaenus, 27

Parurmiatherium, 180, 182

Paschatherium, 13, 17–18, 27

Patagium, 80, 144

Patriotheridomys, 53

Pentalophodont pattern, 51

Peradectes, 9–10, *10,* 24, 25

Peratherium, 25, 69

Percrocuta, 137–38

Percrocutids, 137–38, 162, *163,* 193, 208

Peridyromys, 85, 98

Perissodactyls: archaic, 36, 38, 50; Eocene, 36–38, 46, 50–51, 56, 65; in Messel forest, 46; Miocene, 110–

13, 141–42, 157, 169–70; Oligo-
cene, 69, 72, 92; Paleocene, 23;
Pliocene, 212–13; teeth of, 36
Phacochoerus, 140
Phanourios minor, 261
Phenacodontids, 27, 50
Phenacodus, 27–28, 46, 50, plate 1
Phenacolemur, 25
Phiomia, 106
Phlomis, 230
Pholidocarpus, 43
Phorusracids, 48, 49–50
Photosynthetic pathway, Turolian
stage, 179
Phyllotillon, 110, plate 5
Piezodus, 114
Pigs, 113, 131, 132, 134, 170, *171,*
178, 212
Pikas, 78, 154
Pipe snakes, 43
Pithecanthropus erectus, 246
Pitymys, 251
Plagiolophus, 56, 69, 85
Plagiomene, 34
Platybelodon, 106
Platychoerops, 25
Pleistocene epoch, 241–81; climate
of, 241, 242, 246, 252, 263–64;
early, 241–52; glacial–interglacial
cycles during, 252; Holocene in-
terglacial period, 279–80; late,
263–74; mammals of, 238, 280–81;
megafaunal extinction during,
279–80; middle, 252–63; steppe-
tundra biome in, 263–74, 279
Plesiaceratherium, 142; *P. fahlbuschi,*
111; *P. mirallesi,* 111
Plesiadapiforms, 8, 9, 15, 25, 50
Plesiadapis: Eocene, 25; Paleocene,
15, 17; *P. tricuspidens, 19*
Plesiaddax, 182
Plesiarctomys, 31, 51, 53, 69
Plesictis, 90, 123, plate 3
Plesiogale, 90
Plesiogulo, 162
Plesiomeles, 123
Plesiomeryx, 73
Plesiopliopithecus, 139
Plesispermophilus, 78

Pleuraspidotherium: Paleocene, 13–15,
17–18; *P. aumonieri, 16, 18*
Pliocene epoch, 154, 211–40; climate
of, 211, 218; desert–savanna
boundary in, 230; early, 211–19;
Equus–Mammuth event in, 231–36;
glaciation during, 219–31;
"Golden Age," 219; Middle Plio-
cene Warming in, 229; Pliocene–
Pleistocene boundary in, 238;
Wolf Event in, 236–40, 241
Pliocervus, 205, 212
Pliohippus, 232
Pliohyrax, 157, 187
Pliolophus, 36
Pliopetaurista, 218
Pliopithecids, 139, 164, 175–76
Pliopithecus: Miocene, 138, 164; *P. vin-
doboniensis,* 139
Pliotragus, 226, 238
Plioviverrops: Miocene, 123, 145, 159,
193, 208; Pliocene, 213; *P. orbignyi,*
161
Plithocyon, 119, 145
Poaceae, 230
Pomelomeryx, 97
Pontoceros, 265; *P. ambiguus,* 251
Potamochoerus, 140
Potamotherium, 123; *P. valletoni, 125,*
126
Praeorbulina, 123
Praeovibos: Miocene, 182; Pleistocene,
238, 242, 253, 265; *P. priscus,* 253
Praetiglian stage, 229
Primates: Eocene, 34–36, 45; Mio-
cene, 138–40, 162–69, 178; Oligo-
cene, 105; Pliocene, 216
Primelephas gomphoterioides, 231
Pristichampsus, 20, 43, 48
Proailurus: Miocene, 87, 119; *P. lema-
nensis, 94–95*
"Protoantler," 116
Proboscidean event, 105–10
Proboscideans: in Egypt, 106; Holo-
cene interglacial period, 279; Mio-
cene, 106, 108, 156–57, 178, 179,
205; Pleistocene, 253–54; Plio-
cene, 211, 232; primitive, 106
Procapreolus, 185

Procervulus, 116, 134, 141, plate 5, plate 6, plate 7
Proconsul, 139, 149
Procynocephalus, 216
Procyonids, 123
Prodeinotherium, 108; *P. bavaricum*, 108–9
Prodremotherium, 91
Progiraffa, 115
Progiraffinae, 115
Prognathism, 274
Progonomys, 170
Prolagus, 114
Prolatidens, 10
Prolibytherium, 116, 131
Prolimnocyon, 29–30
Promephitis, 162
Promimomys, 219
Propachynolophus, 37–38, 46
Propaleotherium: Eocene, 46; *P. hassiacum*, 39, 46; *P. parvulum*, 46
Propaleotherium hassiacum, 39
Propliopithecus, 138
Propotamochoerus: Miocene, 140, 157, 159, 170, 205; Pliocene, 212, 221; *P. palaeochoerus*, 205; *P. provincialis*, 205
Proputorius, 123
Prosansanosmilus, 99; *P. peregrinus*, 145
Prosantorhinus, 113, 141; *P. germanicus*, 113
Prosimians, 35
Prostrepsiceros, 180, 205
Protaceratherium, 90, 96, 111; *P. minutum*, 96
Protadelomys, 51
Protapirus, 91
Protatera, 218
Protictitherium: Miocene, 123, 137, 145, 159, 191; *P. crassum*, 160, 161, 191
Protoadapis, 35
Protodichobune, 42
Protoryx, 182, 197, plate 11
Protragocerus, 140, 170
Protungulatum, 7
Proviverra, 30, 46, plate 1
Pseudaelurus: Miocene, 87, 119, 122, 145, 156, 196, plate 6; *P. lorteti*,

120, 121, *121*, 156; *P. quadridentatus*, 121–22, 145, 156, 174; *P. tournauensis*, 119–21, 145, 156
Pseudamphimeryx, 63
Pseudarctos bavaricus, 105, 159, 173
Pseudobassaris, 88, 90
Pseudocricetodon, 80, 98
Pseudocyon: *P. caucasicus*, 104–5; *P. sansaniensis*, 104
Pseudodryomys, 98
Pseudoloris, 53, 69
Pseudoltinomys: Eocene, 51, 53; Oligocene, 78, *82–83*; *P. gaillardi*, 79
Pseudoparamys, 31
Pseudosciurids, 51, 53
Pseudotheridomys, 114
Pterodon: Eocene, 56, 65; Oligocene, 69; *P. dasyuroides*, 58
Ptilodonts, 4, 24
Ptilodus, 4
Puma, 223
Purgatorius, 7, 8

Quercylurus, 81, 83
Quercysorex, 78
Quercytherium, 56

Raccoon dogs, 207, 213, 236
Racoons, 123
Raghatherium, 59
Ramys, 202
Rangifer tarandus, 265, *271*
Rats, 280
Rattus: *R. norvegicus*, 280–81; *R. rattus*, 280
Ratufa, 31
Red deer, 247, *271*
Reduncines, 208
Reedbucks, 208
Reindeer, 265, *271*
Remiculus, 17
Reptiles, 218, 234
Rhagapodemus, 218
Rhinoceroses, 69; African, 91; Asiatic, 91; black, 147, 212; horned, 72, 185, 265; hornless, *158;* Miocene, early, 96, *96*, 111; Miocene, late, 162, 169, 170, 178, 179, 185; Miocene, middle, 134, 141–42, 147;

Oligocene, 90; Pleistocene, 247, *249;* Pliocene, 221–22; two-horned, 221–22; white, *191,* 212, *249;* woolly, *249,* 265, *268, 271,* 279

Rhinocerotids: archaic, 38, 40; Miocene, 111, 113, 157; Oligocene, 70, 91

Rhinocerotines: Miocene, 90, 142, 185; Oligocene, 91, 92

Rhodanomys, 85, 98, 114

Rifian Corridor, 204

Riss-Würm interglacial phase, 264

Ritteneria, 99, 114

Robiacina, 61

Rodents: archaic, 31; endemism among, 262; Eocene, 31, 45, 51, 53; gliding, 135, 142; Holocene interglacial period, 279; in Messel forest, 45; microtine, 218, 251; Miocene, early, 98, 110, 114, 115; Miocene, late, 170, 178, 187, 195; Miocene, middle, 135, 142, 144, 145; Oligocene, 69, 78, 80, 85, 98; Paleocene, 21, 24; Pleistocene, 241, 262; Pliocene, 218–19, 240; sigmodont, 218–19

Ronzotherium, 70, 72, 90

Rotundomys, 173, 187, 218

Ruminants: Eocene, 64; Miocene, 97, 113, 129, 152; Oligocene, 73, 91; Pliocene, 212

Rupicaprine bovids, 238, *240*

Rupricapra rupricapra, 265

Ruscinomys, 135, 218, 220

Sabadellictis, 162

Saber-toothed cats, 81, 154, *192,* 253

Saharan–Sahelian boundary, 230

Saiga targarica, 265

Samotherium, 182

Samotragus, 180, 205

Sanitheres, 134

Sansanosmilus: Miocene, 154–55, 156, plate 8; *S. jourdani,* 145; *S. palmidens,* 145, *146–47, 148*

Saturninia, 53

"Savanna-mosaic" chronofauna, 128

Saxonella, 17

Schizochoerus, 134, 140, 170, 174

Schizotherinae, 72, 110, 142

Schizotheriodes, 38

Schizotherium, 72, 111

Schlossericyon viverroides, 123

Sciuridae, 78

Sciurids, 98, 142

Sciurodon, 78

Sciuroides, 53

Sclerophyllous vegetation, 173

Selenodont dentition, 14, 59, 61

Selenodontia, 47

"Selenolophodont" pattern, 37

Semigenetta: S. elegans, 123; *S. ripolli,* 159

Shamolagus, 78

Sheep, 250, 253

Shrews, 52, 78, 179, 262

Shungura Formation, 231

Sigmodont pattern, 170, 173, 218

Sima de los Huesos (Spain), hominid fossils at, 258, *259*

Simamphicyon helveticus, 52

Sivachoerus, 132

Sivaonyx, 162

Sivapithecus, 147, 164, 175, 179

Sivapterodon, 122

Sivatherines, 154, 182, *182–83*

Siwaliks (Pakistan), 179, 208

Skunks, 123, 162

Smilodon, 174, 226, 253

Soergelia: Pleistocene, 242, 244, 253; *S. elisabethae,* 253; *S. minor,* 242, 252

Soricids, 52

Soricomorphs, 52

Southern Hemisphere, glaciation in, 229

Spaniella, 27

Squirrels, 31, 78, 98, 142, 170, 218, 220; gliding, *121;* ground, 135, 187

Stehlinoceros, 141, plate 8

Stenailurus, 187

Steneofiber, 78

Stenogale, 87

Stenonine horses, *233,* 247, 253

Stenoplesictis, 83

Stephanomys, 218

Stephanorhinus: Pliocene, 221–22; *S. elatus,* 221; *S. etruscus,* 221–22, 231, 247, 251, 252, 256, plate 16; *S. hemitoechus,* 247, 249, 253; *S. hundseimensis,* 247; *S. jeanvireti,* 221–22; *S. kirchbergensis,* 247, 253; *S. megarhinus,* 212; *S. miguelcrusafonti,* 212, 213; *S. pikermiensis,* 185; *S. schleiermarcheri,* 170, 185, 212
Steppe-tundra biome, 263–74, 279
Stertomys, 199, 202
Sub-Paratethyan Province, 128
Subchilotherium, 162
Suevosciurus, Eocene, 53
Suids: Miocene, late, 157, 159, 170, 171, 198, 205; Miocene, middle, 131, 132, 134, 140; Pliocene, 212
Suiforms, 47
Suinae, 197
Suoids, 96
Sus, 140; *S. arvernensis,* 212, 234; *S. scrofa,* 247, 253, 256, plate 15; *S. strozii,* 252
Susliks, 265

Tadarida stehlini, 99, 100, 101, 102
Taeniodonts, 21
Taeniodus, 78
Taeniolabids, 4
Tahr, 242
Takin, 180
Talpids, 52
Talps, 78
Tamandua, 48
Tapirids, 91
Tapiriscus pannonicus, 169
Tapirs: Miocene, 95, 157, 169, 205; Pleistocene, 280; Pliocene, 213, 234
Tapirulus, 61, 72
Tapirus, plate 12; *T. arvernensis,* 234; *T. priscus,* 157
Tarsal anatomy, *Hyopsodus,* 27
Tarsiers, 35
Tarsiforms, 34
Taucanamo, 113, 134, 140
Taurotragus, 251
Taxodon, 123

Tcherskia triton, 80
Tectonic features. *See* Landmasses
Teilhardina, 35
Teleoceratheres, 91
Teleoceratines, 90, 92, 96, 113, 141, 212
Temperature: Eocene, 23; Miocene, 93; Paleocene, 3
Terricola, 251
Teruel Basin (Spain), 179
Teruelia, 115
Testudo, 234
Tethyerians, 105–6
Tethys Sea, 68
Tethytheres, 27
Tethytragus, 130, 140, 182
Tetraconodontines, 132, 134, 140, 170
Tetracus, 78
Tetralophodon, 107, 145, 156, 205, 206, plate 9, plate 11
Teutomanis, plate 6
Thalassictis: Miocene, 145, 159, 162, 174, 193, 208; *T. montadai,* 145; *T. robusta,* 145
Thaumastocyon, 137; *T. dirus,* 137, 159, 173
Theridomorphs: Eocene, 51, 53, 65; Miocene, 98, 114; Oligocene, 85
Theridomyids, 51, 53, 78, 85
Theridomys, 53, 78, plate 3
Theropithecus oswaldi, 251
Thyrrenotragus, 197, 199
Tiglian phase, 241
Tillodonts, 21, 28, 50
Titanoides, 20
Trade winds, Pliocene glaciation period, 230
Tragelaphine bovids, 251
Tragelaphus, 251
Tragoportax: Miocene, 170, 180, 205, plate 10; *T. gaudryi,* 170, 172
Tragulids, 114–15, 159, 179, 180
Tragulus, 64; *T. meminna,* 113
Treposciurus, 53
Triceromeryx, 117, plate 7; *T. conquensis,* 117, 117; *T. pachecoi,* 117, 135
Tricuspiodon, 13
Trilophodon, 108

Trilophomys, 218
Trochictis, 123, 162
Tubulidentata, 97
Tundra. *See* Steppe-tundra biome
Turcocerus, 130
Turgai Strait, 7, 68, 77
Turiacemas, plate 11
Turtles, giant, 218, 234
Tusco-Sardinian island, 193–203
Tusks, 106, 247, 264
Tyrrhenoglis, 261

Udabnopithecus, 147, 193
Umbrotherium, 197
Ungulates, 9; archaic, 7; Eocene, 36–
    42; Holocene interglacial period,
    279; limbs of, 36, 37; Miocene,
    157, 187; Oligocene, 69–77, 90–
    92; Pleistocene, 242, 244; Plio-
    cene, 226, 238
Urmiatherium, 180, 182
Ursavus: Miocene, 119, 145, 173, 193;
    U. brevirhinus, 119, 145, 159; *U. de-
    pereti*, 174; *U. primaevus*, 145, 159
Ursids: Miocene, early, 119; Miocene,
    late, 159, 173–74, 197, 199, 207;
    Miocene, middle, 145; Oligocene,
    88; Pliocene, 214, 216; teeth of,
    119
Ursinae, 173
Ursus: Pleistocene, middle, 256; *U.
    arctos*, 207, 216, 250; *U. deningeri*,
    250, 252, 270; *U. etruscus*, 222,
    226, 231, 244, 250, 251; *U. mini-
    mus*, 214, 216, 222; *U. spelaeus*,
    250, 270, *272*, 273, *273*; *U. thibe-
    tanus*, 216

Vallès-Penedès Basin (Spain), 93, 177
Vallesian climax, 159–62
Vallesian Crisis, 169–78, 193
Vallesian hominoid radiation, 162–69
Vasseuromys, 202
Vegetation. *See* Flora
Vienna Basin (Austria), 123
Viretailurus schaubi, 223, 226, 251
Viverra, 208
Viverrids, 83, 123, 159, 208
Voles, 80, 251
Vulpes, 207, 208, 250; *V. alopecoides*,
    213, 222, 236; *V. praeglacialis*, 236,
    250, 252, 256

Warthogs, 140
West Antarctic Ice Sheet, 229
Wildebeests (gnus), 197
Wolf Event, Pliocene, 236–40, 241
Wolverines, 123, 162, *271*
Wolves, *234*, 236, 250, *271*
Woodlands, in Greek-Iranian Prov-
    ince, 128–38
Wormwood, 230

Xenarthrales, 48
Xenocyon: X. falconeri, 236–37; *X.
    lycaonoides*, 251
Xenohyus, 96
Xiphodon, 63, plate 2
Xiphodontids, 61, 63, 72

Ysengrinia, 104; *Y. valentiana*, 137

Zebras, 233
Zho-Khou-Dien (China), 253
Zramys, 135
Zygolophodon, 108